Springer INdAM Series

Volume 1

For further volumes:
www.springer.com/series/10283

Graziano Gentili · Irene Sabadini ·
Michael Shapiro · Franciscus Sommen ·
Daniele C. Struppa
Editors

Advances in Hypercomplex Analysis

Springer

Editors
Graziano Gentili
Department of Mathematics "U. Dini"
University of Florence
Florence, Italy

Francis Sommen
Department of Mathematical Analysis
University of Ghent
Ghent, Belgium

Irene Sabadini
Dipartimento di Matematica
Politecnico di Milano
Milan, Italy

Daniele C. Struppa
Schmid College of Science and Technology
Chapman University
Orange, CA, USA

Michael Shapiro
Department of Mathematics
E.S.F.M. del I.P.N.
Mexico D.F., Mexico

ISSN 2281-518X ISSN 2281-5198 (electronic)
Springer INdAM Series
ISBN 978-88-470-2444-1 ISBN 978-88-470-2445-8 (eBook)
DOI 10.1007/978-88-470-2445-8
Springer Milan Heidelberg New York Dordrecht London

Library of Congress Control Number: 2012943164

Mathematics Subject Classification: 30G35, 46F15, 47A60, 30C80

Printed on acid-free paper

Springer is part of Springer Science+Business Media (www.springer.com)

Preface

The main goal of a workshop is to bring together international leading specialists in a certain field of research, as well as young mathematicians interested in the subject, with the idea of presenting and discussing recent results, analyzing new trends and techniques in the area and, in general, of promoting scientific collaboration. This has also been the aim of the INdAM Workshop on "Different Notions of Regularity for Functions of a Quaternionic Variable" that took place in Rome, 13–17 September 2010, at the Istituto Nazionale di Alta Matematica "Francesco Severi".

We believe that a good way to have a perception of the impact of a meeting in the interested scientific community is to trace research results that are follow-ups to the lectures and discussions which took place during the event. This is the reason why this volume collects recent and new results in the field of Quaternionic and Clifford Analysis, which are connected to the topic covered during the INdAM Workshop in Rome, 2010.

It is a great pleasure to thank the Istituto Nazionale di Alta Matematica "Francesco Severi", its President Vincenzo Ancona and its entire staff, that made the workshop and this follow-ups volume possible. Our most sincere gratitude goes to Francesca Bonadei, whose highly professional help permitted this publication.

Florence, Italy	Graziano Gentili
Milan, Italy	Irene Sabadini
Mexico D.F., Mexico	Michael Shapiro
Ghent, Belgium	Franciscus Sommen
Orange, CA, USA	Daniele C. Struppa

Contents

Regular vs. Classical Möbius Transformations of the Quaternionic Unit Ball

Cinzia Bisi and Caterina Stoppato

Abstract The regular fractional transformations of the extended quaternionic space have been recently introduced as variants of the classical linear fractional transformations. These variants have the advantage of being included in the class of slice regular functions, introduced by Gentili and Struppa in 2006, so that they can be studied with the useful tools available in this theory. We first consider their general properties, then focus on the regular Möbius transformations of the quaternionic unit ball $\mathbb{B}$, comparing the latter with their classical analogs. In particular we study the relation between the regular Möbius transformations and the Poincaré metric of $\mathbb{B}$, which is preserved by the classical Möbius transformations. Furthermore, we announce a result that is a quaternionic analog of the Schwarz-Pick lemma.

1 Classical Möbius Transformations and Poincaré Distance on the Quaternionic Unit Ball

A classical topic in quaternionic analysis is the study of Möbius transformations. It is well known that the set of *linear fractional transformations* of the extended quaternionic space $\mathbb{H} \cup \{\infty\} \cong \mathbb{HP}^1$

$$\mathbb{G} = \left\{ g(q) = (aq+b) \cdot (cq+d)^{-1} \;\middle|\; \begin{bmatrix} a & b \\ c & d \end{bmatrix} \in GL(2,\mathbb{H}) \right\} \tag{1}$$

Both authors are partially supported by GNSAGA INdAM and by FIRB "Geometria Differenziale Complessa e Dinamica Olomorfa". The second author is also partially supported by FSE and by Regione Lombardia.

C. Bisi
Department of Mathematics and Informatics, University of Ferrara, Via Machiavelli 35, 44121 Ferrara, Italy
e-mail: cinzia.bisi@unife.it

C. Stoppato (✉)
Department of Mathematics "F. Enriques", University of Milan, Via Saldini 50, 20133 Milan, Italy
e-mail: caterina.stoppato@unimi.it

G. Gentili et al. (eds.), *Advances in Hypercomplex Analysis*, Springer INdAM Series 1,
DOI 10.1007/978-88-470-2445-8_1,

is a group with respect to the composition operation. We recall that $GL(2,\mathbb{H})$ denotes the group of 2×2 invertible quaternionic matrices, and that $SL(2,\mathbb{H})$ denotes the subgroup of those such matrices which have unit Dieudonné determinant (for details, see [3] and references therein). It is known in literature that $\mathbb{G}$ is isomorphic to $PSL(2,\mathbb{H}) = SL(2,\mathbb{H})/\{\pm \mathrm{Id}\}$ and that all of its elements are conformal maps. Among the works that treat this matter, even in the more general context of Clifford Algebras, let us mention [1, 10, 19]. The alternative representation

$$\mathbb{G} = \left\{ F_A(q) = (qc+d)^{-1}\cdot(qa+b) \;\middle|\; A = \begin{bmatrix} a & c \\ b & d \end{bmatrix} \in GL(2,\mathbb{H}) \right\} \tag{2}$$

is also possible, and the anti-homomorphism $GL(2,\mathbb{H}) \to \mathbb{G}$ mapping A to F_A induces an anti-isomorphism between $PSL(2,\mathbb{H})$ and $\mathbb{G}$. The group $\mathbb{G}$ is generated by the following four types of transformations:

(i) $L_1(q) = q + b$, $b \in \mathbb{H}$;
(ii) $L_2(q) = q\cdot a$, $a \in \mathbb{H}$, $|a| = 1$;
(iii) $L_3(q) = r\cdot q = q\cdot r$, $r \in \mathbb{R}^+ \setminus \{0\}$;
(iv) $L_4(q) = q^{-1}$.

Moreover, if $\mathscr{S}_i$ is the family of all real i-dimensional spheres, if $\mathscr{P}_i$ is the family of all real i-dimensional affine subspaces of $\mathbb{H}$ and if $\mathscr{F}_i = \mathscr{S}_i \cup \mathscr{P}_i$ then $\mathbb{G}$ maps elements of $\mathscr{F}_i$ onto elements of $\mathscr{F}_i$, for $i = 3, 2, 1$. At this regard, see [20]; detailed proofs of all these facts can be found in [3].

The subgroup $\mathbb{M} \leq \mathbb{G}$ of *(classical) Möbius transformations* mapping the quaternionic open unit ball

$$\mathbb{B} = \{ q \in \mathbb{H} \mid |q| < 1 \}$$

onto itself has also been studied in detail. Let us denote $H = \begin{bmatrix} 1 & 0 \\ 0 & -1 \end{bmatrix}$ and

$$Sp(1,1) = \{ C \in GL(2,\mathbb{H}) \mid \overline{C}^t HC = H \} \subset SL(2,\mathbb{H}).$$

Theorem 1 *An element $g \in \mathbb{G}$ is a classical Möbius transformation of $\mathbb{B}$ if and only if $g(q) = (qc+d)^{-1}\cdot(qa+b)$ with $\begin{bmatrix} a & c \\ b & d \end{bmatrix} \in Sp(1,1)$. This is equivalent to*

$$g(q) = v^{-1}(1 - q\bar{q}_0)^{-1}(q - q_0)u$$

for some $u, v \in \partial\mathbb{B}$, $q_0 \in \mathbb{B}$.

For a proof, see [3]. Hence, $\mathbb{M}$ is anti-isomorphic to $Sp(1,1)/\{\pm \mathrm{Id}\}$. Since $\mathbb{G}$ leaves invariant the family $\mathscr{F}_1$ of circles and affine lines of $\mathbb{H}$, and since the elements of $\mathbb{G}$ are conformal, the group $\mathbb{M}$ of classical Möbius transformations of $\mathbb{B}$ preserves the following class of curves.

Definition 1 If $q_1 \neq q_2 \in \mathbb{B}$ are $\mathbb{R}$-linearly dependent, then the diameter of $\mathbb{B}$ through q_1, q_2 is called the *non-Euclidean line* through q_1 and q_2. Otherwise, the *non-Euclidean line* through q_1 and q_2 is defined as the unique circle through q_1, q_2 that intersects $\partial\mathbb{B} = \mathbb{S}^3$ orthogonally.

Theorem 2 *The formula*

$$\delta_{\mathbb{B}}(q_1, q_2) = \frac{1}{2}\log\left(\frac{1+|1-q_1\bar{q}_2|^{-1}|q_1-q_2|}{1-|1-q_1\bar{q}_2|^{-1}|q_1-q_2|}\right) \tag{3}$$

(for $q_1, q_2 \in \mathbb{B}$) defines a distance that has the non-Euclidean lines as its geodesics. The elements of $\mathbb{M}$ and the map $q \mapsto \bar{q}$ are all isometries for $\delta_{\mathbb{B}}$.

We refer the reader to [2]; a detailed presentation can be found in [3].

So far, we mentioned properties of the classical Möbius transformations that are completely analogous to the complex case. However, the analogy fails on one crucial point. The group $\mathbb{M}$ is not included in the best known analog of the class of holomorphic functions: the set of Fueter regular functions, i.e., the kernel of $\frac{\partial}{\partial x_0} + i\frac{\partial}{\partial x_1} + j\frac{\partial}{\partial x_2} + k\frac{\partial}{\partial x_3}$ (see [18]). For instance, none of the rotations $q \mapsto aq$ with $a \in \mathbb{H}$, $a \neq 0$ is Fueter regular, nor are any of the transformations listed as (i), (ii), (iii), (iv) in our previous discussion. The variant of the Fueter class considered in [11], defined as the kernel of $\frac{\partial}{\partial x_0} + i\frac{\partial}{\partial x_1} + j\frac{\partial}{\partial x_2} - k\frac{\partial}{\partial x_3}$, includes part of the group, for instance the rotations $q \mapsto qb$ for $b \in \mathbb{H}$, $b \neq 0$, but not all of it (for instance, the left multiplication $q \mapsto kq$ by the imaginary unit k is not in the kernel, nor is $q \mapsto q^{-1}$).

A more recent theory of quaternionic functions, introduced in [7, 8], has proven to be more comprehensive. The theory is based on the next definition.

Definition 2 Let Ω be a domain in $\mathbb{H}$ and let $f : \Omega \to \mathbb{H}$ be a function. For all $I \in \mathbb{S} = \{q \in \mathbb{H} \mid q^2 = -1\}$, let us denote $L_I = \mathbb{R} + I\mathbb{R}$, $\Omega_I = \Omega \cap L_I$ and $f_I = f_{|\Omega_I}$. The function f is called (Cullen or) *slice regular* if, for all $I \in \mathbb{S}$, the restriction f_I is real differentiable and the function $\bar{\partial}_I f : \Omega_I \to \mathbb{H}$ defined by

$$\bar{\partial}_I f(x+Iy) = \frac{1}{2}\left(\frac{\partial}{\partial x} + I\frac{\partial}{\partial y}\right) f_I(x+Iy)$$

vanishes identically.

By direct computation, the class of slice regular functions includes all of the generators we listed as (i), (ii), (iii), (iv). It does not contain the whole group $\mathbb{G}$ (nor its subgroup $\mathbb{M}$), because composition does not, in general, preserve slice regularity. However, [17] introduced the new classes of (*slice*) *regular fractional transformations* and (*slice*) *regular Möbius transformations* of $\mathbb{B}$, which are nicely related to the classical linear fractional transformations. They are presented in detail in Sects. 2 and 3.

One of the purposes of the present paper is, in fact, to compare the slice regular fractional transformations with the classical ones. Furthermore, we take a first glance at the role played by slice regular Möbius transformations in the geometry of $\mathbb{B}$. In Sect. 4, we undertake a first study of their differential properties: we prove that they are not, in general, conformal, and that they do not preserve the Poincaré distance $\delta_{\mathbb{B}}$. In Sect. 5, we announce a result of [4]: a quaternionic analog of the Schwarz-Pick lemma, which discloses the possibility of using slice regular functions in the study of the intrinsic geometry of $\mathbb{B}$.

2 Regular Fractional Transformations

This section surveys the algebraic structure of slice regular functions, and its application to the construction of regular fractional transformations. From now on, we will omit the term 'slice' and refer to these functions as regular, *tout court*. Since we will be interested only in regular functions on Euclidean balls

$$B(0, R) = \{q \in \mathbb{H} \mid |q| < R\},$$

or on the whole space $\mathbb{H} = B(0, +\infty)$, we will follow the presentation of [6, 16]. However, we point out that many of the results we are about to mention have been generalized to a larger class of domains in [5].

Theorem 3 *Fix R with $0 < R \leq +\infty$ and let $\mathscr{D}_R$ be the set of regular functions $f : B(0, R) \to \mathbb{H}$. Then $\mathscr{D}_R$ coincides with the set of quaternionic power series $f(q) = \sum_{n\in\mathbb{N}} q^n a_n$ (with $a_n \in \mathbb{H}$) converging in $B(0, R)$. Moreover, defining the* regular multiplication $*$ *by the formula*

$$\left(\sum_{n\in\mathbb{N}} q^n a_n\right) * \left(\sum_{n\in\mathbb{N}} q^n b_n\right) = \sum_{n\in\mathbb{N}} q^n \sum_{k=0}^{n} a_k b_{n-k}, \tag{4}$$

we conclude that $\mathscr{D}_R$ is an associative real algebra with respect to $+, *$.

The ring $\mathscr{D}_R$ admits a classical ring of quotients

$$\mathscr{L}_R = \{f^{-*} * g \mid f, g \in \mathscr{D}_R,\ f \not\equiv 0\}.$$

In order to introduce it, we begin with the following definition.

Definition 3 Let $f(q) = \sum_{n\in\mathbb{N}} q^n a_n$ be a regular function on an open ball $B = B(0, R)$. The regular conjugate of f, $f^c : B \to \mathbb{H}$, is defined as $f^c(q) = \sum_{n\in\mathbb{N}} q^n \bar{a}_n$ and the symmetrization of f, as $f^s = f * f^c = f^c * f$.

Notice that $f^s(q) = \sum_{n\in\mathbb{N}} q^n r_n$ with $r_n = \sum_{k=0}^{n} a_k \bar{a}_{n-k} \in \mathbb{R}$. Moreover, the zero sets of f^c and f^s have been fully characterized.

Theorem 4 *Let f be a regular function on $B = B(0, R)$. For all $x, y \in \mathbb{R}$ with $x + y\mathbb{S} \subseteq B$, the regular conjugate f^c has as many zeros as f in $x + y\mathbb{S}$. Moreover, the zero set of the symmetrization f^s is the union of the $x + y\mathbb{S}$ on which f has a zero.*

We are now ready for the definition of regular quotient. We denote by

$$\mathscr{Z}_h = \{q \in B \mid h(q) = 0\}$$

the zero set of a function h.

Definition 4 Let $f, g : B = B(0, R) \to \mathbb{H}$ be regular functions. The *left regular quotient* of f and g is the function $f^{-*} * g$ defined in $B \setminus \mathscr{Z}_{f^s}$ by

$$f^{-*} * g(q) = f^s(q)^{-1} f^c * g(q). \tag{5}$$

Moreover, the *regular reciprocal* of f is the function $f^{-*} = f^{-*} * 1$.

Left regular quotients proved to be regular in their domains of definition. If we set $(f^{-*} * g) * (h^{-*} * k) = (f^s h^s)^{-1} f^c * g * h^c * k$ then $(\mathscr{L}_R, +, *)$ is a division algebra over $\mathbb{R}$ and it is the classical ring of quotients of $(\mathscr{D}_R, +, *)$ (see [14]). In particular, $\mathscr{L}_R$ coincides with the set of *right regular quotients*

$$g * h^{-*}(q) = h^s(q)^{-1} g * h^c(q).$$

The definition of regular conjugation and symmetrization is extended to $\mathscr{L}_R$ setting $(f^{-*} * g)^c = g^c * (f^c)^{-*}$ and $(f^{-*} * g)^s(q) = f^s(q)^{-1} g^s(q)$. Furthermore, the following relation between the left regular quotient $f^{-*} * g(q)$ and the quotient $f(q)^{-1} g(q)$ holds.

Theorem 5 *Let f, g be regular functions on $B = B(0, R)$. Then*

$$f * g(q) = \begin{cases} 0 & \text{if } f(q) = 0, \\ f(q) g(f(q)^{-1} q f(q)) & \text{otherwise.} \end{cases} \tag{6}$$

Setting $T_f(q) = f^c(q)^{-1} q f^c(q)$ for all $q \in B \setminus \mathscr{Z}_{f^s} \subseteq B \setminus \mathscr{Z}_{f^c}$,

$$f^{-*} * g(q) = f\big(T_f(q)\big)^{-1} g\big(T_f(q)\big), \tag{7}$$

for all $q \in B \setminus \mathscr{Z}_{f^s}$. For all $x, y \in \mathbb{R}$ with $x + y\mathbb{S} \subset B \setminus \mathscr{Z}_{f^s}$, the function T_f maps $x + y\mathbb{S}$ to itself (in particular $T_f(x) = x$ for all $x \in \mathbb{R}$). Furthermore, T_f is a diffeomorphism from $B \setminus \mathscr{Z}_{f^s}$ onto itself, with inverse T_{f^c}.

We point out that, so far, no simple result relating $g * h^{-*}(q)$ to $g(q)h(q)^{-1}$ is known.

This machinery allowed the introduction in [17] of regular analogs of linear fractional transformations, and of Möbius transformations of $\mathbb{B}$. To each $A = \left[\begin{smallmatrix} a & c \\ b & d \end{smallmatrix}\right] \in GL(2, \mathbb{H})$ we can associate the *regular fractional transformation*

$$\mathscr{F}_A(q) = (qc + d)^{-*} * (qa + b).$$

By the formula $(qc + d)^{-*} * (qa + b)$ we denote the aforementioned left regular quotient $f^{-*} * g$ of $f(q) = qc + d$ and $g(q) = qa + b$. We denote the 2×2 identity matrix as Id.

Theorem 6 *Choose $R > 0$ and consider the ring of quotients of regular quaternionic functions in $B(0, R)$, denoted by $\mathscr{L}_R$. Setting*

$$f.A = (fc + d)^{-*} * (fa + b) \tag{8}$$

for all $f \in \mathscr{L}_R$ and for all $A = \left[\begin{smallmatrix} a & c \\ b & d \end{smallmatrix}\right] \in GL(2, \mathbb{H})$ defines a right action of $GL(2, \mathbb{H})$ on $\mathscr{L}_R$. A left action of $GL(2, \mathbb{H})$ on $\mathscr{L}_R$ is defined setting

$$A^t.f = (a * f + b) * (c * f + d)^{-*}. \tag{9}$$

The stabilizer of any element of $\mathcal{L}_R$ with respect to either action includes the normal subgroup $N = \{t \cdot \mathrm{Id} \mid t \in \mathbb{R} \setminus \{0\}\} \trianglelefteq GL(2,\mathbb{H})$. Both actions are faithful when reduced to $PSL(2,\mathbb{H}) = GL(2,\mathbb{H})/N$.

The statements concerning the right action are proven in [17], the others can be similarly derived. The two actions coincide in one special case.

Proposition 1 *For all Hermitian matrices $A = \left[\begin{smallmatrix} a & \bar{b} \\ b & d \end{smallmatrix}\right]$ with $a, d \in \mathbb{R}$, $b \in \mathbb{H}$,*

$$f.A = (f\bar{b}+d)^{-*} * (fa+b) = (a*f+b)*(\bar{b}*f+d)^{-*} = A^t.f.$$

Proof We observe that

$$(f\bar{b}+d)^{-*} * (fa+b) = (a*f+b)*(\bar{b}*f+d)^{-*}$$

if, and only if,

$$(fa+b)*(\bar{b}*f+d) = (f\bar{b}+d)*(a*f+b),$$

which is equivalent to

$$af*\bar{b}*f+|b|^2 f+adf+db = af*\bar{b}*f+adf+|b|^2 f+db. \qquad \square$$

In general, a more subtle relation holds between the two actions.

Remark 1 For all $A \in GL(2,\mathbb{H})$ and for all $f \in \mathcal{L}_R$, by direct computation

$$(f.A)^c = \bar{A}^t.f^c.$$

As a consequence, if $A \in GL(2,\mathbb{H})$ is Hermitian then $(f.A)^c = f^c.\bar{A}$.

Interestingly, neither action is free, not even when reduced to $PSL(2,\mathbb{H})$. Indeed, the stabilizer of the identity function with respect to either action of $GL(2,\mathbb{H})$ equals

$$\{c \cdot \mathrm{Id} \mid c \in \mathbb{H} \setminus \{0\}\},$$

a subgroup of $GL(2,\mathbb{H})$ that strictly includes N and is not normal. As a consequence, the set of regular fractional transformations

$$\mathfrak{G} = \{\mathcal{F}_A \mid A \in GL(2,\mathbb{R})\} = \{\mathcal{F}_A \mid A \in SL(2,\mathbb{R})\},$$

which is the orbit of the identity function $id = \mathcal{F}_{\mathrm{Id}}$ under the right action of $GL(2,\mathbb{H})$ on $\mathcal{L}_\infty$, does not inherit a group structure from $GL(2,\mathbb{H})$.

Lemma 1 *The set $\mathfrak{G}$ of regular fractional transformations is also the orbit of id with respect to the left action of $GL(2,\mathbb{H})$ on $\mathcal{L}_\infty$.*

Proof Let $A = \left[\begin{smallmatrix} a & c \\ b & d \end{smallmatrix}\right] \in GL(2,\mathbb{H})$, and let us prove that $\mathcal{F}_A = id.A$ can also be expressed as $C.id$ for some $C \in GL(2,\mathbb{H})$. If $c = 0$ then

$$\mathcal{F}_A(q) = d^{-1} * (qa+b) = \left(d^{-1}a\right) * q + d^{-1}b,$$

else

$$\mathscr{F}_A(q) = \mathscr{F}_{c^{-1}A}(q) = (q-p)^{-*} * (q\alpha + \beta) = \left[(q-p)^s\right]^{-1}(q-\bar{p}) * (q\alpha+\beta)$$

for some $p, \alpha, \beta \in \mathbb{H}$. If $p = x + Iy$ then there exists $\tilde{p} \in x + y\mathbb{S}$ and $\gamma, \delta \in \mathbb{H}$ such that $(q-\bar{p}) * (q\alpha+\beta) = (q\gamma+\delta) * (q-\tilde{p})$; additionally, $(q-p)^s = (q-\tilde{p})^s$ (see [6] for details). Hence,

$$\begin{aligned}\mathscr{F}_A(q) &= \left[(q-\tilde{p})^s\right]^{-1}(q\gamma+\delta) * (q-\tilde{p}) \\ &= (q\gamma+\delta) * (q-\bar{\tilde{p}})^{-*} = (\gamma * q + \delta) * (q - \bar{\tilde{p}})^{-*},\end{aligned}$$

which is of the desired form. Similar manipulations prove that for all $C \in GL(2, \mathbb{H})$, the function $C.id$ equals $\mathscr{F}_A = id.A$ for some $A \in GL(2, \mathbb{H})$. □

We now state an immediate consequence of the previous lemma and of Remark 1.

Remark 2 The set $\mathfrak{G}$ of regular fractional transformations is preserved by regular conjugation.

3 Regular Möbius Transformations of $\mathbb{B}$

The regular fractional transformations that map the open quaternionic unit ball $\mathbb{B}$ onto itself, called *regular Möbius transformations of* $\mathbb{B}$, are characterized by two results of [17].

Theorem 7 *For all $A \in SL(2, \mathbb{H})$, the regular fractional transformation $\mathscr{F}_A$ maps $\mathbb{B}$ onto itself if and only if $A \in Sp(1,1)$, if and only if there exist (unique) $u \in \partial\mathbb{B}$, $a \in \mathbb{B}$ such that*

$$\mathscr{F}_A(q) = (1 - q\bar{a})^{-*} * (q-a)u. \tag{10}$$

In particular, the set $\mathfrak{M} = \{f \in \mathfrak{G} \mid f(\mathbb{B}) = \mathbb{B}\}$ of the regular Möbius transformations of $\mathbb{B}$ is the orbit of the identity function under the right action of $Sp(1,1)$.

Theorem 8 *The class of regular bijective functions $f : \mathbb{B} \to \mathbb{B}$ coincides with the class $\mathfrak{M}$ of regular Möbius transformations of $\mathbb{B}$.*

As a consequence, the right action of $Sp(1,1)$ preserves the class of regular bijective functions from $\mathbb{B}$ onto itself. We now study, more in general, the effect of the actions of $Sp(1,1)$ on the class

$$\mathfrak{Reg}(\mathbb{B}, \mathbb{B}) = \{f : \mathbb{B} \to \mathbb{B} \mid f \text{ is regular}\}.$$

Proposition 2 *If $f \in \mathfrak{Reg}(\mathbb{B}, \mathbb{B})$ then for all $a \in \mathbb{B}$*

$$(1 - f\bar{a})^{-*} * (f-a) = (f-a) * (1 - \bar{a} * f)^{-*}. \tag{11}$$

Moreover, the left and right actions of $Sp(1,1)$ preserve $\mathfrak{Reg}(\mathbb{B}, \mathbb{B})$.

Proof The fact that $(1 - f\bar{a})^{-*} * (f - a) = (f - a) * (1 - \bar{a} * f)^{-*}$ is a consequence of Proposition 1.

Let us turn to the second statement, proving that for all $a \in \mathbb{B}$, $u, v \in \partial\mathbb{B}$, the function

$$\tilde{f} = v^{-*} * (1 - f\bar{a})^{-*} * (f - a)u = (v - f\bar{a}v)^{-*} * (f - a)u$$

is in $\mathfrak{Reg}(\mathbb{B}, \mathbb{B})$. The fact that for all $a \in \mathbb{B}$, $u, v \in \partial\mathbb{B}$ the function $u * (f - a) * (1 - \bar{a} * f)^{-*} * v^{-*}$ belongs to $\mathfrak{Reg}(\mathbb{B}, \mathbb{B})$ will then follow from the equality just proven.

The function $\tilde{f}$ is regular in $\mathbb{B}$ since $h = v - f\bar{a}v$ has no zero in $\mathbb{B}$ (as a consequence of the fact that $|a| < 1$, $|f| < 1$ and $|v| = 1$). Furthermore,

$$\begin{aligned}\tilde{f} &= \left(v - (f \circ T_h)\bar{a}v\right)^{-1}(f \circ T_h - a)u \\ &= v^{-1}\left(1 - (f \circ T_h)\bar{a}\right)^{-1}(f \circ T_h - a)u = v^{-1}(M_a \circ f \circ T_h)u\end{aligned}$$

where T_h and $M_a(q) = (1 - q\bar{a})^{-1}(q - a)$ map $\mathbb{B}$ to itself bijectively and $u, v \in \partial\mathbb{B}$. Hence, $\tilde{f} = v^{-*} * (1 - f\bar{a})^{-*} * (f - a) * u \in \mathfrak{Reg}(\mathbb{B}, \mathbb{B})$, as desired. □

As a byproduct, we obtain that the orbit of the identity function under the left action of $Sp(1, 1)$ equals $\mathfrak{M}$.

Proposition 3 *If $f \in \mathfrak{Reg}(\mathbb{B}, \mathbb{B})$ then its regular conjugate f^c belongs to $\mathfrak{Reg}(\mathbb{B}, \mathbb{B})$ as well. Furthermore, f^c is bijective (hence an element of $\mathfrak{M}$) if and only if f is.*

Proof Suppose $f^c(p) = a \in \mathbb{H} \setminus \mathbb{B}$ for some $p = x + Iy \in \mathbb{B}$. Then p is a zero of the regular function $f^c - a$. By Theorem 4, there exists $\tilde{p} \in x + y\mathbb{S} \subset \mathbb{B}$ such that $(f^c - a)^c = f - \bar{a}$ vanishes at $\tilde{p}$. Hence, $f(\mathbb{B})$ includes $\bar{a} \in \mathbb{H} \setminus \mathbb{B}$, a contradiction with the hypothesis $f(\mathbb{B}) \subseteq \mathbb{B}$.

As for the second statement, f^c is bijective if and only if, for all $a \in \mathbb{B}$, there exists a unique $p \in \mathbb{B}$ such that $f^c(p) = a$. Reasoning as above, we conclude that this happens if and only if for all $a \in \mathbb{B}$, there exists a unique $\tilde{p} \in \mathbb{B}$ such that $f(\tilde{p}) = \bar{a}$. This is equivalent to the bijectivity of f. □

4 Differential and Metric Properties of Regular Möbius Transformations

The present section is concerned with two natural questions:

(a) whether the regular Möbius transformations are conformal (as the classical Möbius transformations);
(b) whether they preserve the quaternionic Poincaré distance defined on $\mathbb{B}$ by formula (3).

For a complete description of the Poincaré metric, see [3]. In order to answer question (a), we will compute for a regular Möbius transformation the series development introduced by the following result of [15]. Let us set

$$U(x_0+y_0\mathbb{S},r)=\{q\in\mathbb{H}\mid |(q-x_0)^2+y_0^2|<r^2\}$$

for all $x_0, y_0 \in \mathbb{R}$, $r>0$.

Theorem 9 *Let f be a regular function on $\Omega = B(0,R)$, and let $U(x_0+y_0\mathbb{S},r)\subseteq\Omega$. Then for each $q_0\in x_0+y_0\mathbb{S}$ there exists $\{A_n\}_{n\in\mathbb{N}}\subset\mathbb{H}$ such that*

$$f(q)=\sum_{n\in\mathbb{N}}\left[(q-x_0)^2+y_0^2\right]^n\left[A_{2n}+(q-q_0)A_{2n+1}\right] \tag{12}$$

for all $q\in U(x_0+y_0\mathbb{S},r)$. As a consequence,

$$\frac{\partial f}{\partial v}(q_0)=\lim_{t\to 0}\frac{f(q_0+tv)-f(q_0)}{t}=vA_1+(q_0v-v\bar{q}_0)A_2 \tag{13}$$

for all $v\in T_{q_0}\Omega\cong\mathbb{H}$.

If $q_0\in L_I$ and if we split the tangent space $T_{q_0}\Omega\cong\mathbb{H}$ as $\mathbb{H}=L_I\oplus L_I^{\perp}$, then the differential of f at q_0 acts by right multiplication by A_1 on $L_I^{\perp}$ and by right multiplication by $A_1+2\,\mathrm{Im}(q_0)A_2$ on L_I. Furthermore, if for all $q_0\in\Omega$ the *differential quotient* $R_{q_0}f$ is defined as

$$R_{q_0}f(q)=(q-q_0)^{-*}*\big(f(q)-f(q_0)\big)$$

then the coefficients of (12) are computed as $A_{2n}=(R_{\bar{q}_0}R_{q_0})^n f(q_0)$ and $A_{2n+1}=R_{q_0}(R_{\bar{q}_0}R_{q_0})^n f(\bar{q}_0)$.

Let us recall the definition of the *Cullen derivative* $\partial_c f$, given in [8] as

$$\partial_c f(x+Iy)=\frac{1}{2}\left(\frac{\partial}{\partial x}-I\frac{\partial}{\partial y}\right)f(x+Iy) \tag{14}$$

for $I\in\mathbb{S}$, $x,y\in\mathbb{R}$, as well as the definition of the *spherical derivative*

$$\partial_s f(q)=\big(2\,\mathrm{Im}(q)\big)^{-1}\big(f(q)-f(\bar{q})\big) \tag{15}$$

given in [9]. We can make the following observation.

Remark 3 If f is a regular function on $B(0,R)$ and if (12) holds then $\partial_c f(q_0)=R_{q_0}f(q_0)=A_1+2\,\mathrm{Im}(q_0)A_2$ and $\partial_s f(q_0)=R_{q_0}f(\bar{q}_0)=A_1$.

In the case of the regular Möbius transformation

$$\mathcal{M}_{q_0}(q)=(1-q\bar{q}_0)^{-*}*(q-q_0)=(q-q_0)*(1-q\bar{q}_0)^{-*},$$

clearly $R_{q_0}\mathcal{M}_{q_0}(q)=(1-q\bar{q}_0)^{-*}$, so that we easily compute the coefficients A_n.

Proposition 4 *Let $q_0=x_0+y_0I\in\mathbb{B}$. Then the expansion* (12) *of $\mathcal{M}_{q_0}$ at q_0 has coefficients*

$$A_{2n-1} = \frac{\bar{q}_0^{2n-2}}{(1-|q_0|^2)^{n-1}(1-\bar{q}_0^2)^n}, \tag{16}$$

$$A_{2n} = \frac{\bar{q}_0^{2n-1}}{(1-|q_0|^2)^{n}(1-\bar{q}_0^2)^n} \tag{17}$$

for all $n \geq 1$. As a consequence, for all $v \in \mathbb{H}$,

$$\frac{\partial \mathcal{M}_{q_0}}{\partial v}(q_0) = v\big(1-\bar{q}_0^2\big)^{-1} + (q_0 v - v\bar{q}_0)\frac{\bar{q}_0}{(1-|q_0|^2)(1-\bar{q}_0^2)}. \tag{18}$$

Proof We have already observed that $R_{q_0}\mathcal{M}_{q_0}(q) = (1-q\bar{q}_0)^{-*}$, so that $A_1 = R_{q_0}\mathcal{M}_{q_0}(\bar{q}_0) = (1-\bar{q}_0^2)^{-1}$. Moreover,

$$\begin{aligned} R_{\bar{q}_0}R_{q_0}\mathcal{M}_{q_0}(q) &= (q-\bar{q}_0)^{-*} * \big[R_{q_0}\mathcal{M}_{q_0}(q) - A_1\big] \\ &= (q-\bar{q}_0)^{-*} * \big[(1-q\bar{q}_0)^{-*} - A_1\big] \\ &= (1-q\bar{q}_0)^{-*} * (q-\bar{q}_0)^{-*} * \big[\big(1-\bar{q}_0^2\big) - (1-q\bar{q}_0)\big]A_1 \\ &= (1-q\bar{q}_0)^{-*}\bar{q}_0 A_1. \end{aligned}$$

The thesis follows by induction, proving that for all $n \geq 1$

$$\begin{aligned} (R_{\bar{q}_0}R_{q_0})^n\mathcal{M}_{q_0}(q) &= (1-q\bar{q}_0)^{-*}\bar{q}_0 A_{2n-1}, \\ R_{q_0}(R_{\bar{q}_0}R_{q_0})^n\mathcal{M}_{q_0}(q) &= (1-q\bar{q}_0)^{-*}\bar{q}_0 A_{2n} \end{aligned}$$

by means of similar computations.

Formula (18) is a direct application of the previous formula (13) and of the above computations. □

We are now in a position to answer question (a).

Remark 4 For each $q_0 = x_0 + Iy_0 \in \mathbb{B}\setminus\mathbb{R}$, the differential of $\mathcal{M}_{q_0}$ at q_0 acts by right multiplication by $\partial_c\mathcal{M}_{q_0}(q_0) = (1-|q_0|^2)^{-1}$ on L_I and by right multiplication by $\partial_s\mathcal{M}_{q_0}(q_0) = (1-\bar{q}_0^2)^{-1}$ on $L_I^\perp$. Since L_I, $L_I^\perp$ are both invariant and $\partial_c\mathcal{M}_{q_0}(q_0)$, $\partial_s\mathcal{M}_{q_0}(q_0)$ have different moduli, $\mathcal{M}_{q_0}$ is not conformal.

We now turn our attention to question (b): whether or not regular Möbius transformations preserve the quaternionic Poincaré metric on $\mathbb{B}$ described in Sect. 1 and in [3]. We recall that this metric was constructed to be preserved by the classical (nonregular) Möbius transformations of $\mathbb{B}$. Thanks to Theorem 5, we observe what follows.

Remark 5 If $\mathcal{M}_{q_0}(q) = (1-q\bar{q}_0)^{-*} * (q-q_0)$ and $M_{q_0}(q) = (1-q\bar{q}_0)^{-1}(q-q_0)$ then

$$\mathcal{M}_{q_0}(q) = M_{q_0}\big(T(q)\big) \tag{19}$$

where $T(q) = (1-qq_0)^{-1}q(1-qq_0)$ is a diffeomorphism of $\mathbb{B}$ with inverse function $T^{-1}(q) = (1-q\bar{q}_0)^{-1}q(1-q\bar{q}_0)$.

Thus, a generic regular Möbius transformation of $\mathbb{B}$

$$q \mapsto \mathcal{M}_{q_0}(q)u = M_{q_0}\big(T(q)\big)u$$

(with $u \in \partial\mathbb{B}$) is an isometry if and only if T is. An example shows that this is not the case.

Example 1 Let $q_0 = \frac{I_0}{2}$ for some $I_0 \in \mathbb{S}$. Then $T(q) = (1 - qq_0)^{-1}q(1 - qq_0)$ is not an isometry for the Poincaré metric defined by formula (3). Indeed, if $J_0 \in \mathbb{S}$, $J_0 \perp I_0$ then setting $q_1 = \frac{J_0}{2}$ we have

$$\delta_{\mathbb{B}}(q_0, q_1) > \delta_{\mathbb{B}}\big(T(q_0), T(q_1)\big)$$

since

$$\frac{|q_1 - q_0|^2}{|1 - q_1\bar{q}_0|^2} = \frac{|2J_0 - 2I_0|^2}{|4 + J_0 \cdot I_0|^2} = \frac{8}{17}$$

while, computing $T(q_0) = q_0 = \frac{I_0}{2}$ and $T(q_1) = \frac{8I_0+15J_0}{34}$, we conclude that

$$\frac{|T(q_1) - q_0|^2}{|1 - T(q_1)\bar{q}_0|^2} = 4\frac{|-3I_0 + 5J_0|^2}{|20 - 5\, I_0 \cdot J_0|^2} = \frac{8}{25}.$$

Thus, the regular Möbius transformations do not have a definite behavior with respect to $\delta_{\mathbb{B}}$: we have seen that T (hence $\mathcal{M}_{q_0}$) is not an isometry, nor a dilation; the same computation shows that $T^{-1}(q) = (1 - q\bar{q}_0)^{-1}q(1 - q\bar{q}_0)$ (hence $\mathcal{M}_{\bar{q}_0}$) is not a contraction.

The previous discussion proves that the study of regular Möbius transformations cannot be framed into the classical study of $\mathbb{B}$, and that it requires further research. On the other hand, the theory of regular functions provides working tools that were not available for the classical Möbius transformations. These tools led us in [4] to an analog of the Schwarz-Pick lemma, which we will present in the next section.

5 The Schwarz-Pick Lemma for Regular Functions

In the complex case, holomorphic functions play a crucial role in the study of the intrinsic geometry of the unit disc $\Delta = \{z \in \mathbb{C} \mid |z| < 1\}$ thanks to the Schwarz-Pick lemma [12, 13].

Theorem 10 *Let $f : \Delta \to \Delta$ be a holomorphic function and let $z_0 \in \Delta$. Then*

$$\left|\frac{f(z) - f(z_0)}{1 - \overline{f(z_0)}f(z)}\right| \leq \left|\frac{z - z_0}{1 - \bar{z}_0 z}\right|, \tag{20}$$

for all $z \in \Delta$ and

$$\frac{|f'(z_0)|}{1 - |f(z_0)|^2} \leq \frac{1}{1 - |z_0|^2}. \tag{21}$$

All inequalities are strict for $z \in \Delta \setminus \{z_0\}$, unless f is a Möbius transformation.

This is exactly the type of tool that is not available, in the quaternionic case, for the classical Möbius transformations. To the contrary, an analog of the Schwarz-Pick lemma is proven in [4] for quaternionic regular functions. To present it, we begin with a result concerning the special case of a function $f \in \mathfrak{Reg}(\mathbb{B}, \mathbb{B})$ having a zero.

Theorem 11 *If $f : \mathbb{B} \to \mathbb{B}$ is regular and if $f(q_0) = 0$ for some $q_0 \in \mathbb{B}$, then*

$$\left|\mathcal{M}_{q_0}^{-*} * f(q)\right| \leq 1 \tag{22}$$

for all $q \in \mathbb{B}$. The inequality is strict, unless $\mathcal{M}_{q_0}^{-*} * f(q) \equiv u$ for some $u \in \partial\mathbb{B}$.

A useful property of the moduli of regular products is also proven in [4]:

Lemma 2 *Let $f, g, h : B(0, R) \to \mathbb{H}$ be regular functions. If $|f| \leq |g|$ then $|h * f| \leq |h * g|$. Moreover, if $|f| < |g|$ then $|h * f| < |h * g|$ in $B \setminus \mathcal{Z}_h$.*

The property above allows us to derive from Theorem 11 the perfect analog of the Schwarz-Pick lemma in the special case $f(q_0) = 0$. We recall that $\partial_c f$ denotes the Cullen derivative of f, defined by formula (14), while $\partial_s f$ denotes the spherical derivative, defined by formula (15).

Corollary 1 *If $f : \mathbb{B} \to \mathbb{B}$ is regular and if $f(q_0) = 0$ for some $q_0 \in \mathbb{B}$ then*

$$\left|f(q)\right| \leq \left|\mathcal{M}_{q_0}(q)\right| \tag{23}$$

for all $q \in \mathbb{B}$. The inequality is strict at all $q \in \mathbb{B} \setminus \{q_0\}$, unless there exists $u \in \partial\mathbb{B}$ such that $f(q) = \mathcal{M}_{q_0}(q) \cdot u$ at all $q \in \mathbb{B}$. Moreover, $|R_{q_0} f(q)| \leq |(1 - q\bar{q}_0)^{-*}|$ in $\mathbb{B}$ and in particular

$$\left|\partial_c f(q_0)\right| \leq \frac{1}{1 - |q_0|^2}, \tag{24}$$

$$\left|\partial_s f(q_0)\right| \leq \frac{1}{|1 - \bar{q}_0^2|}. \tag{25}$$

These inequalities are strict, unless $f(q) = \mathcal{M}_{q_0}(q) \cdot u$ for some $u \in \partial\mathbb{B}$.

Finally, the desired result is obtained in full generality.

Theorem 12 (Schwarz-Pick lemma) *Let $f : \mathbb{B} \to \mathbb{B}$ be a regular function and let $q_0 \in \mathbb{B}$. Then*

$$\left|(f(q) - f(q_0)) * \left(1 - \overline{f(q_0)} * f(q)\right)^{-*}\right| \leq \left|(q - q_0) * (1 - \bar{q}_0 * q)^{-*}\right|, \tag{26}$$

$$\left|R_{q_0} f(q) * \left(1 - \overline{f(q_0)} * f(q)\right)^{-*}\right| \leq \left|(1 - \bar{q}_0 * q)^{-*}\right| \tag{27}$$

in $\mathbb{B}$. In particular,

$$\left|\partial_c f * \left(1 - \overline{f(q_0)} * f(q)\right)^{-*}\right|_{|q_0} \leq \frac{1}{1 - |q_0|^2}, \tag{28}$$

$$\frac{|\partial_s f(q_0)|}{|1 - f^s(q_0)|} \leq \frac{1}{|1 - \bar{q}_0^2|}. \tag{29}$$

If f is a regular Möbius transformation of $\mathbb{B}$ then equality holds in (26), (27) *for all $q \in \mathbb{B}$, and in* (28), (29). *Else, all the aforementioned inequalities are strict* (*except for* (26) *at q_0, which reduces to* $0 \leq 0$).

This promising result makes it reasonable to expect that regular functions play an important role in the intrinsic geometry of the quaternionic unit ball. Therefore, it encourages to continue the study of regular Möbius transformations and of their differential or metric properties.

References

1. Ahlfors, L.V.: Möbius transformations in $\mathbf{R}^n$ expressed through 2×2 matrices of Clifford numbers. Complex Var. Theory Appl. **5**(2–4), 215–224 (1986)
2. Ahlfors, L.V.: Cross-ratios and Schwarzian derivatives in $\mathbf{R}^n$. In: Complex Analysis, pp. 1–15. Birkhäuser, Basel (1988)
3. Bisi, C., Gentili, G.: Möbius transformations and the Poincaré distance in the quaternionic setting. Indiana Univ. Math. J. **58**(6), 2729–2764 (2009)
4. Bisi, C., Stoppato, C.: The Schwarz-Pick lemma for slice regular functions. Indiana Univ. Math. J. (2012). doi:10.1512/iumj.2012.61.5076
5. Colombo, F., Gentili, G., Sabadini, I., Struppa, D.: Extension results for slice regular functions of a quaternionic variable. Adv. Math. **222**, 1793–1808 (2009)
6. Gentili, G., Stoppato, C.: Zeros of regular functions and polynomials of a quaternionic variable. Mich. Math. J. **56**(3), 655–667 (2008)
7. Gentili, G., Struppa, D.C.: A new approach to Cullen-regular functions of a quaternionic variable. C. R. Math. Acad. Sci. Paris **342**(10), 741–744 (2006)
8. Gentili, G., Struppa, D.C.: A new theory of regular functions of a quaternionic variable. Adv. Math. **216**(1), 279–301 (2007)
9. Ghiloni, R., Perotti, A.: Slice regular functions on real alternative algebras. Adv. Math. **226**(2), 1662–1691 (2011)
10. Maass, H.: Automorphe Funktionen von meheren Veränderlichen und Dirichletsche Reihen. Abh. Math. Semin. Univ. Hamb. **16**, 72–100 (1949)
11. Perotti, A.: Regular quaternionic functions and conformal mappings. CUBO **11**(1), 123–143 (2009)
12. Pick, G.: Über eine Eigenschaft der konformen Abbildung kreisförmiger Bereiche. Math. Ann. **77**(1), 1–6 (1915)
13. Pick, G.: Über die Beschränkungen analytischer Funktionen, welche durch vorgegebene Funktionswerte bewirkt werden. Math. Ann. **77**(1), 7–23 (1915)
14. Rowen, L.H.: Ring Theory, student edition. Academic Press, Boston (1991)
15. Stoppato, C.: A new series expansion for slice regular functions. Adv. Math. **231**, 1401–1416 (2012)
16. Stoppato, C.: Poles of regular quaternionic functions. Complex Var. Elliptic Equ. **54**(11), 1001–1018 (2009)
17. Stoppato, C.: Regular Moebius transformations of the space of quaternions. Ann. Glob. Anal. Geom. **39**(4), 387–401 (2011)
18. Sudbery, A.: Quaternionic analysis. Math. Proc. Camb. Philos. Soc. **85**(2), 199–224 (1979)
19. Vahlen, K.T.: Ueber Bewegungen und complexe Zahlen. Math. Ann. **55**(4), 585–593 (1902)
20. Wilker, J.B.: The quaternion formalism for Möbius groups in four or fewer dimensions. Linear Algebra Appl. **190**, 99–136 (1993)

Distributional Boundary Values of Harmonic Potentials in Euclidean Half-Space as Fundamental Solutions of Convolution Operators in Clifford Analysis

Fred Brackx, Hendrik De Bie, and Hennie De Schepper

Abstract In the framework of Clifford analysis, a chain of harmonic and monogenic potentials in the upper half of Euclidean space $\mathbb{R}^{m+1}$ was recently constructed, including a higher dimensional analogue of the logarithmic function in the complex plane. In this construction the distributional limits of these potentials at the boundary $\mathbb{R}^m$ are crucial. The remarkable relationship between these distributional boundary values and four basic pseudodifferential operators linked with the Dirac and Laplace operators is studied.

1 Introduction

In a recent paper [7] a generalization to Euclidean upper half-space $\mathbb{R}^{m+1}_+$ was constructed of the logarithmic function $\ln z$ which is holomorphic in the upper half of the complex plane. This construction was carried out in the framework of Clifford analysis, where the functions under consideration take their values in the universal Clifford algebra $\mathbb{R}_{0,m+1}$ constructed over the Euclidean space $\mathbb{R}^{m+1}$, equipped with a quadratic form of signature $(0, m+1)$. The concept of a higher dimensional holomorphic function, mostly called monogenic function, is expressed by means of a generalized Cauchy-Riemann operator, which is a combination of the derivative with respect to one of the real variables, say x_0, and the so-called Dirac operator $\underline{\partial}$ in the remaining real variables $(x_1, x_2, \ldots, x_m)$. The generalized Cauchy-Riemann operator D and its Clifford algebra conjugate $\overline{D}$ linearize the Laplace operator, whence Clifford analysis may be seen as a refinement of harmonic analysis.

The starting point of the construction of a higher dimensional monogenic logarithmic function, was the fundamental solution of the generalized Cauchy-Riemann

F. Brackx · H. De Bie · H. De Schepper (✉)
Clifford Research Group, Department of Mathematical Analysis, Ghent University, Galglaan 2 (S22), 9000 Ghent, Belgium
e-mail: hds@cage.ugent.be

F. Brackx
e-mail: fb@cage.ugent.be

H. De Bie
e-mail: hdebie@cage.ugent.be

G. Gentili et al. (eds.), *Advances in Hypercomplex Analysis*, Springer INdAM Series 1,
DOI 10.1007/978-88-470-2445-8_2, © Springer-Verlag Italia 2013

operator D, also called Cauchy kernel, and its relation to the Poisson kernel and its harmonic conjugate in $\mathbb{R}^{m+1}_+$. We then proceeded by induction in two directions, *downstream* by differentiation and *upstream* by primitivation, yielding a doubly infinite chain of monogenic, and thus harmonic, potentials. This chain mimics the well-known sequence of holomorphic potentials in $\mathbb{C}_+$ (see e.g. [12]):

$$\frac{1}{k!}z^k\left[\ln z - \left(1+\frac{1}{2}+\cdots+\frac{1}{k}\right)\right]$$
$$\to\cdots\to z(\ln z-1)\to \ln z \xrightarrow{\frac{d}{dz}} \frac{1}{z}\to -\frac{1}{z^2}\to\cdots\to(-1)^{k-1}\frac{(k-1)!}{z^k}.$$

Identifying the boundary of upper half-space with $\mathbb{R}^m \cong \{(x_0,\underline{x})\in\mathbb{R}^{m+1} : x_0=0\}$, the distributional limits for $x_0\to 0+$ of those potentials were computed. They split up into two classes of distributions, which are linked by the Hilbert transform, one scalar-valued, the second one Clifford vector-valued. They form two of the four families of Clifford distributions which were thoroughly studied in a series of papers, see [4–6] and the references therein.

These distributional boundary values are really fundamental, since not only they are used in the definition of the harmonic and monogenic potentials, but they also uniquely determine the conjugate harmonic potentials obtained by primitivation, thanks to the simple, but crucial, fact that a monogenic function in $\mathbb{R}^{m+1}_+$ vanishing at the boundary $\mathbb{R}^m$ indeed is zero. Whence the need to predict the distributional boundary values when constructing the, at that moment unknown, upstream potentials. To that end the distributional boundary values have to be identified in some way ab initio, which is the aim of the present paper. It is shown that half of them may be recovered as fundamental solutions of specific powers of the Dirac operator, and also half of them, but not the missing ones, as fundamental solutions of specific powers of the Laplace operator. By introducing two new pseudodifferential operators, related to the complex powers of the Dirac and Laplace operators, the whole double infinite set of distributional boundary values may now be identified as fundamental solutions of the four operators. As a remarkable sign of symmetry, the distributional boundary values also can serve as convolution kernels for the corresponding pseudodifferential operators of the same kind but with opposite exponent.

The organization of the paper is as follows. To make the paper self-contained we recall in Sect. 2 the basics of Clifford algebra and Clifford analysis and in Sect. 3 the main results of [7] on the conjugate harmonic and monogenic potentials in upper half-space $\mathbb{R}^{m+1}_+$. The four pseudodifferential operators needed for recovering all the distributional boundary values of these harmonic potentials as fundamental solutions, are studied in four consecutive sections. Sections 4 and 6 are devoted to the complex powers of the Dirac and Laplace operator respectively and their fundamental solutions. In Sects. 5 and 7 the two new operators, also depending on a complex parameter, and their fundamental solutions are studied. Section 8 contains some conclusions.

2 Basics of Clifford Analysis

Clifford analysis (see e.g. [3]) is a function theory which offers a natural and elegant generalization to higher dimension of holomorphic functions in the complex plane and refines harmonic analysis. Let $(e_0, e_1, \ldots, e_m)$ be the canonical orthonormal basis of Euclidean space $\mathbb{R}^{m+1}$ equipped with a quadratic form of signature $(0, m+1)$. Then the non-commutative multiplication in the universal real Clifford algebra $\mathbb{R}_{0,m+1}$ is governed by the rule

$$e_\alpha e_\beta + e_\beta e_\alpha = -2\delta_{\alpha\beta}, \quad \alpha, \beta = 0, 1, \ldots, m$$

whence $\mathbb{R}_{0,m+1}$ is generated additively by the elements $e_A = e_{j_1} \ldots e_{j_h}$, where $A = \{j_1, \ldots, j_h\} \subset \{0, \ldots, m\}$, with $0 \leq j_1 < j_2 < \cdots < j_h \leq m$, and $e_\emptyset = 1$. For an account on Clifford algebra we refer to e.g. [13].

We identify the point $(x_0, x_1, \ldots, x_m) \in \mathbb{R}^{m+1}$ with the Clifford-vector variable

$$x = x_0 e_0 + x_1 e_1 + \cdots + x_m e_m = x_0 e_0 + \underline{x}$$

and the point $(x_1, \ldots, x_m) \in \mathbb{R}^m$ with the Clifford-vector variable $\underline{x}$. The introduction of spherical co-ordinates $\underline{x} = r\underline{\omega}$, $r = |\underline{x}|$, $\underline{\omega} \in S^{m-1}$, gives rise to the Clifford-vector valued locally integrable function $\underline{\omega}$, which is to be seen as the higher dimensional analogue of the *signum*-distribution on the real line; we will encounter $\underline{\omega}$ as one of the distributions discussed below.

At the heart of Clifford analysis lies the so-called Dirac operator

$$\partial = \partial_{x_0} e_0 + \partial_{x_1} e_1 + \cdots + \partial_{x_m} e_m = \partial_{x_0} e_0 + \underline{\partial}$$

which squares to the negative Laplace operator: $\partial^2 = -\Delta_{m+1}$, while also $\underline{\partial}^2 = -\Delta_m$. The fundamental solution of the Dirac operator ∂ is given by

$$E_{m+1}(x) = -\frac{1}{\sigma_{m+1}} \frac{x}{|x|^{m+1}}$$

where $\sigma_{m+1} = \frac{2\pi^{\frac{m+1}{2}}}{\Gamma(\frac{m+1}{2})}$ stands for the area of the unit sphere S^m in $\mathbb{R}^{m+1}$. We also introduce the generalized Cauchy-Riemann operator

$$D = \frac{1}{2}\overline{e_0}\partial = \frac{1}{2}(\partial_{x_0} + \overline{e_0}\,\underline{\partial})$$

which, together with its Clifford algebra conjugate $\overline{D} = \frac{1}{2}(\partial_{x_0} - \overline{e_0}\,\underline{\partial})$, also decomposes the Laplace operator: $D\overline{D} = \overline{D}D = \frac{1}{4}\Delta_{m+1}$.

A continuously differentiable function $F(x)$, defined in an open region $\Omega \subset \mathbb{R}^{m+1}$ and taking values in the Clifford algebra $\mathbb{R}_{0,m+1}$, is called (left-)monogenic if it satisfies in Ω the equation $DF = 0$, which is equivalent with $\partial F = 0$.

Singling out the basis vector e_0, we can decompose the real Clifford algebra $\mathbb{R}_{0,m+1}$ in terms of the Clifford algebra $\mathbb{R}_{0,m}$ as $\mathbb{R}_{0,m+1} = \mathbb{R}_{0,m} \oplus \overline{e_0}\,\mathbb{R}_{0,m}$. Similarly we decompose the considered functions as

$$F(x_0, \underline{x}) = F_1(x_0, \underline{x}) + \overline{e_0} F_2(x_0, \underline{x})$$

where F_1 and F_2 take their values in the Clifford algebra $\mathbb{R}_{0,m}$; mimicking functions of a complex variable, we will call F_1 the *real* part and F_2 the *imaginary* part of the function F.

We will extensively use two families of distributions in $\mathbb{R}^m$, which have been thoroughly studied in [4–6]. The first family $\mathscr{T} = \{T_\lambda : \lambda \in \mathbb{C}\}$ is very classical. It consists of the radial distributions

$$T_\lambda = \mathrm{Fp}\, r^\lambda = \mathrm{Fp}\big(x_1^2 + \cdots + x_m^2\big)^{\frac{\lambda}{2}}$$

their action on a test function $\phi \in \mathscr{S}(\mathbb{R}^m)$ being given by

$$\langle T_\lambda, \phi\rangle = \sigma_m \big\langle \mathrm{Fp}\, r_+^\mu, \Sigma^{(0)}[\phi]\big\rangle$$

with $\mu = \lambda + m - 1$. In the above expressions $\mathrm{Fp}\, r_+^\mu$ stands for the classical "finite part" distribution on the real r-axis and $\Sigma^{(0)}$ is the scalar valued generalized spherical mean, defined on scalar valued test functions $\phi(\underline{x})$ by

$$\Sigma^{(0)}[\phi] = \frac{1}{\sigma_m} \int_{S^{m-1}} \phi(\underline{x})\, dS(\underline{\omega}).$$

This family $\mathscr{T}$ contains a.o. the fundamental solutions of the natural powers of the Laplace operator. As convolution operators they give rise to the traditional Riesz potentials (see e.g. [11]). The second family $\mathscr{U} = \{U_\lambda : \lambda \in \mathbb{C}\}$ of distributions arises in a natural way by the action of the Dirac operator $\underline{\partial}$ on $\mathscr{T}$. The U_λ-distributions thus are typical Clifford analysis constructs: they are Clifford-vector valued, and they also arise as products of T_λ-distributions with the distribution $\underline{\omega} = \frac{\underline{x}}{|\underline{x}|}$, mentioned above. The action of U_λ on a test function $\phi \in \mathscr{S}(\mathbb{R}^m)$ is given by

$$\langle U_\lambda, \phi\rangle = \sigma_m \big\langle \mathrm{Fp}\, r_+^\mu, \Sigma^{(1)}[\phi]\big\rangle$$

with $\mu = \lambda + m - 1$, and where the Clifford-vector valued generalized spherical mean $\Sigma^{(1)}$ is defined on scalar valued test functions $\phi(\underline{x})$ by

$$\Sigma^{(1)}[\phi] = \frac{1}{\sigma_m} \int_{S^{m-1}} \underline{\omega}\, \phi(\underline{x})\, dS(\underline{\omega}).$$

Typical examples in the $\mathscr{U}$-family are the fundamental solutions of the Dirac operator and of its odd natural powers.

The normalized distributions T_λ^* and U_λ^* arise when removing the singularities of T_λ and U_λ by dividing them by an appropriate Gamma-function showing the same simple poles. The scalar T_λ^*-distributions are defined by

$$\begin{cases} T_\lambda^* = \pi^{\frac{\lambda+m}{2}} \dfrac{T_\lambda}{\Gamma(\frac{\lambda+m}{2})}, & \lambda \neq -m - 2l \\[2ex] T_{-m-2l}^* = \dfrac{\pi^{\frac{m}{2}-l}}{2^{2l}\,\Gamma(\frac{m}{2}+l)} (-\Delta_m)^l \delta(\underline{x}), & l \in \mathbb{N}_0 \end{cases} \tag{1}$$

while the Clifford-vector valued distributions U_λ^* are defined by

$$\begin{cases} U_\lambda^* = \pi^{\frac{\lambda+m+1}{2}} \dfrac{U_\lambda}{\Gamma(\frac{\lambda+m+1}{2})}, & \lambda \neq -m-2l-1 \\ U_{-m-2l-1}^* = -\dfrac{\pi^{\frac{m}{2}-l}}{2^{2l+1}\Gamma(\frac{m}{2}+l+1)} \underline{\partial}^{2l+1}\delta(\underline{x}), & l \in \mathbb{N}_0. \end{cases} \tag{2}$$

The normalized distributions T_λ^* and U_λ^* are holomorphic mappings from $\lambda \in \mathbb{C}$ to the space $\mathscr{S}'(\mathbb{R}^m)$ of tempered distributions. As already mentioned they are intertwined by the action of the Dirac operator; more generally they enjoy the following properties: for all $\lambda \in \mathbb{C}$ one has:

(i) $\underline{x}T_\lambda^* = \frac{\lambda+m}{2\pi}U_{\lambda+1}^*$; $\underline{x}U_\lambda^* = U_\lambda^*\underline{x} = -T_{\lambda+1}^*$;
(ii) $\underline{\partial}T_\lambda^* = \lambda U_{\lambda-1}^*$; $\underline{\partial}U_\lambda^* = U_\lambda^*\underline{\partial} = -2\pi T_{\lambda-1}^*$;
(iii) $\Delta_m T_\lambda^* = 2\pi\lambda T_{\lambda-2}^*$; $\Delta_m U_\lambda^* = 2\pi(\lambda-1)U_{\lambda-2}^*$;
(iv) $r^2 T_\lambda^* = \frac{\lambda+m}{2\pi}T_{\lambda+2}^*$; $r^2 U_\lambda^* = \frac{\lambda+m+1}{2\pi}U_{\lambda+2}^*$.

Of particular importance for the sequel are convolution formulae for the T_λ^*- and U_λ^*-distributions; we list them in the following proposition and refer the reader to [6] for more details. Notice that the convolution of distributions from both families is commutative notwithstanding the Clifford vector character of the U_λ^*-distributions.

Proposition 1

(i) *For all $(\alpha,\beta) \in \mathbb{C}\times\mathbb{C}$ such that $\alpha \neq 2j$, $j \in \mathbb{N}_0$, $\beta \neq 2k$, $k \in \mathbb{N}_0$ and $\alpha+\beta+m \neq 2l$, $l \in \mathbb{N}_0$ the convolution $T_\alpha^* * T_\beta^*$ is the tempered distribution given by*

$$T_\alpha^* * T_\beta^* = \pi^{\frac{m}{2}} \frac{\Gamma(-\frac{\alpha+\beta+m}{2})}{\Gamma(-\frac{\alpha}{2})\Gamma(-\frac{\beta}{2})} T_{\alpha+\beta+m}^*.$$

(ii) *For $(\alpha,\beta) \in \mathbb{C}\times\mathbb{C}$ such that $\alpha \neq 2j+1$, $\beta \neq 2k$, $\alpha+\beta \neq -m+2l+1$, $j,k,l \in \mathbb{N}_0$ one has*

$$U_\alpha^* * T_\beta^* = T_\beta^* * U_\alpha^* = \pi^{\frac{m}{2}} \frac{\Gamma(-\frac{\alpha+\beta+m-1}{2})}{\Gamma(-\frac{\alpha-1}{2})\Gamma(-\frac{\beta}{2})} U_{\alpha+\beta+m}^*.$$

(iii) *For $(\alpha,\beta) \in \mathbb{C}\times\mathbb{C}$ such that $\alpha \neq 2j+1$, $\beta \neq 2k+1$, $\alpha+\beta \neq -m+2l$, $j,k,l \in \mathbb{N}_0$ one has*

$$U_\alpha^* * U_\beta^* = U_\beta^* * U_\alpha^* = \pi^{\frac{m}{2}+1} \frac{\Gamma(-\frac{\alpha+\beta+m}{2})}{\Gamma(\frac{-\alpha+1}{2})\Gamma(\frac{-\beta+1}{2})} T_{\alpha+\beta+m}^*.$$

3 Harmonic and Monogenic Potentials in $\mathbb{R}_+^{m+1}$

In this section we gather the most important results on harmonic and monogenic potentials in upper half-space $\mathbb{R}^{m+1}$, which were established in [7].

The starting point is the Cauchy kernel of Clifford analysis, i.e. the fundamental solution of the generalized Cauchy-Riemann operator D:

$$C_{-1}(x_0,\underline{x}) = \frac{1}{\sigma_{m+1}} \frac{x\overline{e_0}}{|x|^{m+1}} = \frac{1}{\sigma_{m+1}} \frac{x_0 - \overline{e_0}\,\underline{x}}{|x|^{m+1}}$$

which may be decomposed in terms of the traditional Poisson kernels in $\mathbb{R}^{m+1}_+$:

$$C_{-1}(x_0,\underline{x}) = \frac{1}{2}A_{-1}(x_0,\underline{x}) + \frac{1}{2}\overline{e_0}B_{-1}(x_0,\underline{x})$$

where, also mentioning the usual notations, for $x_0 > 0$,

$$A_{-1}(x_0,\underline{x}) = P(x_0,\underline{x}) = \frac{2}{\sigma_{m+1}} \frac{x_0}{|x|^{m+1}}$$

$$B_{-1}(x_0,\underline{x}) = Q(x_0,\underline{x}) = -\frac{2}{\sigma_{m+1}} \frac{\underline{x}}{|x|^{m+1}}$$

Their distributional limits for $x_0 \to 0+$ are given by

$$a_{-1}(\underline{x}) = \lim_{x_0\to 0+} A_{-1}(x_0,\underline{x}) = \delta(\underline{x}) = \frac{2}{\sigma_m} T^*_{-m}$$

$$b_{-1}(\underline{x}) = \lim_{x_0\to 0+} B_{-1}(x_0,\underline{x}) = H(\underline{x}) = -\frac{2}{\sigma_{m+1}} U^*_{-m}$$

where the distribution

$$H(\underline{x}) = -\frac{2}{\sigma_{m+1}} U^*_{-m} = -\frac{2}{\sigma_{m+1}} \,\mathrm{Pv}\, \frac{\underline{x}}{|\underline{x}|^{m+1}}$$

with Pv standing for the "principal value" distribution in $\mathbb{R}^m$, is the convolution kernel of the Hilbert transform $\mathscr{H}$ in $\mathbb{R}^m$ (see e.g. [9]). Note also that both distributional boundary values are linked by this Hilbert transform:

$$\mathscr{H}[a_{-1}] = \mathscr{H}[\delta] = H * \delta = H = b_{-1}$$

$$\mathscr{H}[b_{-1}] = \mathscr{H}[H] = H * H = \delta = a_{-1}$$

since $\mathscr{H}^2 = \mathbf{1}$.

The first in the sequence of so-called *downstream* potentials is the function C_{-2} defined by

$$\overline{D}C_{-1} = C_{-2} = \frac{1}{2}A_{-2} + \frac{1}{2}\overline{e_0}B_{-2}.$$

Clearly it is monogenic in $\mathbb{R}^{m+1}_+$, since $DC_{-2} = D\overline{D}C_{-1} = \frac{1}{4}\Delta_{m+1}C_{-1} = 0$. The definition itself of $C_{-2}(x_0,\underline{x})$ implies that it shows the monogenic potential (or primitive) $C_{-1}(x_0,\underline{x})$ and the conjugate harmonic potentials $A_{-2}(x_0,\underline{x})$ and $\overline{e_0}B_{-2}(x_0,\underline{x})$. The distributional limits for $x_0 \to 0+$ of these harmonic potentials are given by

$$\begin{cases} a_{-2}(\underline{x}) = \lim\limits_{x_0\to 0+} A_{-2}(x_0,\underline{x}) = \dfrac{2}{\sigma_{m+1}}\,\mathrm{Fp}\,\dfrac{1}{|\underline{x}|^{m+1}} = -\dfrac{4\pi}{\sigma_{m+1}} T^*_{-m-1} \\ b_{-2}(\underline{x}) = \lim\limits_{x_0\to 0+} B_{-2}(x_0,\underline{x}) = -\underline{\partial}\delta = \dfrac{2m}{\sigma_m} U^*_{-m-1}. \end{cases}$$

Proceeding in the same manner, the sequence of *downstream* monogenic potentials in $\mathbb{R}_+^{m+1}$ is defined by

$$C_{-k-1} = \overline{D} C_{-k} = \overline{D}^2 C_{-k+1} = \cdots = \overline{D}^k C_{-1}, \quad k = 1, 2, \ldots$$

each monogenic potential decomposing into two conjugate harmonic potentials:

$$C_{-k-1} = \frac{1}{2} A_{-k-1} + \frac{1}{2} \overline{e_0} B_{-k-1}, \quad k = 1, 2, \ldots$$

with, for k odd, say $k = 2\ell - 1$,

$$\begin{cases} A_{-2\ell} = \partial_{x_0}^{2\ell-1} A_{-1} = -\partial_{x_0}^{2\ell-2} \underline{\partial} B_{-1} = \cdots = -\underline{\partial}^{2\ell-1} B_{-1} \\ B_{-2\ell} = \partial_{x_0}^{2\ell-1} B_{-1} = -\partial_{x_0}^{2\ell-2} \underline{\partial} A_{-1} = \cdots = -\underline{\partial}^{2\ell-1} A_{-1} \end{cases}$$

while for k even, say $k = 2\ell$,

$$\begin{cases} A_{-2\ell-1} = \partial_{x_0}^{2\ell} A_{-1} = -\partial_{x_0}^{2\ell-1} \underline{\partial} B_{-1} = \cdots = \underline{\partial}^{2\ell} A_{-1} \\ B_{-2\ell-1} = \partial_{x_0}^{2\ell} B_{-1} = -\partial_{x_0}^{2\ell-1} \underline{\partial} A_{-1} = \cdots = \underline{\partial}^{2\ell} B_{-1} \end{cases}$$

Their distributional limits for $x_0 \to 0+$ are given by

$$\begin{cases} a_{-2\ell} = (-\underline{\partial})^{2\ell-1} H = -2^{2\ell-1} \dfrac{\Gamma(\frac{m+2\ell-1}{2})}{\pi^{\frac{m-2\ell+1}{2}}} T^*_{-m-2\ell+1} \\ \qquad = (-1)^{\ell-1} 2^{\ell-1} (2\ell-1)!! \dfrac{\Gamma(\frac{m+2\ell-1}{2})}{\pi^{\frac{m+1}{2}}} \operatorname{Fp} \dfrac{1}{r^{m+2\ell-1}} \\ b_{-2\ell} = (-\underline{\partial})^{2\ell-1} \delta = 2^{2\ell-1} \dfrac{\Gamma(\frac{m+2\ell}{2})}{\pi^{\frac{m-2\ell+2}{2}}} U^*_{-m-2\ell+1} \end{cases}$$

and

$$\begin{cases} a_{-2\ell-1} = \underline{\partial}^{2\ell} \delta = 2^{2\ell} \dfrac{\Gamma(\frac{m+2\ell}{2})}{\pi^{\frac{m-2\ell}{2}}} T^*_{-m-2\ell} \\ b_{-2\ell-1} = \underline{\partial}^{2\ell} H = -2^{2\ell} \dfrac{\Gamma(\frac{m+2\ell+1}{2})}{\pi^{\frac{m-2\ell+1}{2}}} U^*_{-m-2\ell} \\ \qquad = (-1)^{\ell-1} 2^{\ell} (2\ell-1)!! \dfrac{\Gamma(\frac{m+2\ell+1}{2})}{\pi^{\frac{m+1}{2}}} \operatorname{Fp} \dfrac{1}{r^{m+2\ell}} \omega \end{cases}$$

They show the following properties.

Lemma 1 *One has for* $j, k = 1, 2, \ldots$

(i) $a_{-k} \xrightarrow{-\underline{\partial}} b_{-k-1} \xrightarrow{-\underline{\partial}} a_{-k-2}$;
(ii) $\mathscr{H}[a_{-k}] = b_{-k}$, $\mathscr{H}[b_{-k}] = a_{-k}$;
(iii) $a_{-j} * a_{-k} = a_{-j-k+1}$
$a_{-j} * b_{-k} = b_{-j} * a_{-k} = b_{-j-k+1}$
$b_{-j} * b_{-k} = a_{-j-k+1}$.

Let us have a look at the so-called *upstream* potentials. To start with the fundamental solution of the Laplace operator Δ_{m+1} in $\mathbb{R}^{m+1}$, sometimes called Green's function, and here denoted by $\frac{1}{2}A_0(x_0,\underline{x})$, is given by

$$\frac{1}{2}A_0(x_0,\underline{x}) = -\frac{1}{m-1}\frac{1}{\sigma_{m+1}}\frac{1}{|x|^{m-1}}.$$

Its conjugate harmonic in $\mathbb{R}^{m+1}_+$, in the sense of [3], is

$$B_0(x_0,\underline{x}) = \frac{2}{\sigma_{m+1}}\frac{\underline{x}}{|\underline{x}|^m}F_m\left(\frac{|\underline{x}|}{x_0}\right) \tag{3}$$

where

$$F_m(v) = \int_0^v \frac{\eta^{m-1}}{(1+\eta^2)^{\frac{m+1}{2}}}\,d\eta = \frac{v^m}{m}\,{}_2F_1\left(\frac{m}{2},\frac{m+1}{2};\frac{m}{2}+1;-v^2\right)$$

with ${}_2F_1$ a standard hypergeometric function (see e.g. [10]). Taking into account that

$$F_m(+\infty) = \int_0^{+\infty} \frac{\eta^{m-1}}{(1+\eta^2)^{\frac{m+1}{2}}}\,d\eta = \frac{\sqrt{\pi}}{2}\frac{\Gamma(\frac{m}{2})}{\Gamma(\frac{m+1}{2})}$$

expression (3) leads to the following distributional limit

$$b_0(\underline{x}) = \lim_{x_0\to 0+} B_0(x_0,\underline{x}) = \frac{1}{\sigma_m}\frac{\underline{x}}{|\underline{x}|^m} = \frac{1}{\pi}\frac{1}{\sigma_m}U^*_{-m+1}$$

while $A_0(x_0,\underline{x})$ itself shows the distributional limit

$$a_0(\underline{x}) = \lim_{x_0\to 0+} A_0(x_0,\underline{x}) = -\frac{2}{m-1}\frac{1}{\sigma_{m+1}}\,\mathrm{Fp}\,\frac{1}{|\underline{x}|^{m-1}} = -\frac{2}{m-1}\frac{1}{\sigma_{m+1}}T^*_{-m+1}.$$

It is readily seen that $\overline{D}A_0 = \overline{D}\overline{e_0}B_0 = C_{-1}$. So $A_0(x_0,\underline{x})$ and $\overline{e_0}B_0(x_0,\underline{x})$ are conjugate harmonic potentials (or primitives), with respect to the operator $\overline{D}$, of the Cauchy kernel $C_{-1}(x_0,\underline{x})$ in $\mathbb{R}^{m+1}_+$. Putting $C_0(x_0,\underline{x}) = \frac{1}{2}A_0(x_0,\underline{x}) + \frac{1}{2}\overline{e_0}B_0(x_0,\underline{x})$, it follows that also $\overline{D}C_0(x_0,\underline{x}) = C_{-1}(x_0,\underline{x})$, which implies that $C_0(x_0,\underline{x})$ is a monogenic potential (or primitive) of the Cauchy kernel $C_{-1}(x_0,\underline{x})$ in $\mathbb{R}^{m+1}_+$. Their distributional boundary values are intimately related, as shown in the following lemma.

Lemma 2 *One has*

(i) $-\underline{\partial}a_0 = b_{-1} = H;\ -\underline{\partial}b_0 = a_{-1} = \delta$
(ii) $\mathscr{H}[a_0] = b_0;\ \mathscr{H}[b_0] = a_0$

Remark 1 In the upper half of the complex plane the function $\ln(z)$ is a holomorphic potential (or primitive) of the Cauchy kernel $\frac{1}{z}$ and its real and imaginary components are the fundamental solution $\ln|z|$ of the Laplace operator, and its conjugate harmonic $i\arg(z)$ respectively. By similarity we could say that $C_0(x_0,\underline{x}) = \frac{1}{2}A_0(x_0,\underline{x}) + \frac{1}{2}\overline{e_0}B_0(x_0,\underline{x})$, being a monogenic potential of the Cauchy kernel

$C_{-1}(x_0, \underline{x})$ and the sum of the fundamental solution $A_0(x_0, \underline{x})$ of the Laplace operator and its conjugate harmonic $\overline{e_0}B_0(x_0, \underline{x})$, is a *monogenic logarithmic function* in the upper half-space $\mathbb{R}^{m+1}_+$.

Inspired by the above mentioned properties, the construction of the sequence of *upstream* harmonic and monogenic potentials in $\mathbb{R}^{m+1}_+$ is continued as follows. Putting

$$\begin{cases} A_1(x_0, \underline{x}) = a_0(\cdot) * A_0(x_0, \cdot)(\underline{x}) = b_0(\cdot) * B_0(x_0, \cdot) \\ B_1(x_0, \underline{x}) = a_0(\cdot) * B_0(x_0, \cdot)(\underline{x}) = b_0(\cdot) * A_0(x_0, \cdot) \end{cases}$$

it is verified that $\overline{D}A_{-1} = \overline{D}\,\overline{e_0}B_{-1} = C_0$, whence $A_1(x_0, \underline{x})$ and $B_1(x_0, \underline{x})$ are conjugate harmonic potentials in $\mathbb{R}^{m+1}_+$ of the function $C_0(x_0, \underline{x})$. It then follows at once that

$$C_1(x_0, \underline{x}) = \frac{1}{2}A_1(x_0, \underline{x}) + \frac{1}{2}\overline{e_0}B_1(x_0, \underline{x})$$

is a monogenic potential in $\mathbb{R}^{m+1}_+$ of C_0. The distributional limits for $x_0 \to 0+$ of the conjugate harmonic potentials A_1 and B_1 are given by

$$\begin{cases} a_1(\underline{x}) = \lim\limits_{x_0 \to 0+} A_1(x_0, \underline{x}) = a_0(\cdot) * a_0(\cdot)(\underline{x}) = b_0(\cdot) * b_0(\cdot)(\underline{x}) \\ b_1(\underline{x}) = \lim\limits_{x_0 \to 0+} B_1(x_0, \underline{x}) = a_0(\cdot) * b_0(\cdot)(\underline{x}) = b_0(\cdot) * a_0(\cdot)(\underline{x}). \end{cases}$$

Making use of the calculation rules for the convolution of the T^*- and U^*-distributions (see Sect. 2, Proposition 1), these distributional boundary values are explicitly given by

$$\begin{cases} a_1(\underline{x}) = \dfrac{1}{\pi}\dfrac{1}{\sigma_m}\dfrac{1}{m-2}T^*_{-m+2} = \dfrac{1}{\sigma_m}\dfrac{1}{m-2}\dfrac{1}{|\underline{x}|^{m-2}} \\ b_1(\underline{x}) = -\dfrac{1}{\pi}\dfrac{1}{\sigma_{m+1}}\dfrac{1}{m-1}U^*_{-m+2} = -\dfrac{1}{\sigma_{m+1}}\dfrac{2}{m-1}\dfrac{\underline{x}}{|\underline{x}|^{m-1}} \end{cases}$$

They show the following properties.

Lemma 3

(i) $-\underline{\partial}a_1 = b_0$, $-\underline{\partial}b_1 = a_0$;
(ii) $\mathscr{H}[a_1] = b_1$, $\mathscr{H}[b_1] = a_1$.

The conjugate harmonic potentials $A_1(x_0, \underline{x})$ and $B_1(x_0, \underline{x})$ have been determined explicitly:

$$\begin{cases} A_1(x_0, \underline{x}) = \dfrac{2}{m-1}\dfrac{1}{\sigma_{m+1}}\dfrac{1}{|\underline{x}|^{m-2}}F_{m-2}\left(\dfrac{\underline{x}}{x_0}\right) \\ B_1(x_0, \underline{x}) = \dfrac{2}{\sigma_{m+1}}\dfrac{x_0\underline{x}}{|\underline{x}|^m}F_m\left(\dfrac{|\underline{x}|}{x_0}\right) - \dfrac{2}{\sigma_{m+1}}\dfrac{1}{m-1}\dfrac{\underline{x}}{|\underline{x}|^{m-1}}. \end{cases}$$

Proceeding in a similar way, it is verified that the functions $A_2(x_0,\underline{x})$ and $B_2(x_0,\underline{x})$ defined by

$$\begin{cases} A_2(x_0,\underline{x}) = a_0(\cdot) * A_1(x_0,\cdot)(\underline{x}) = b_0(\cdot) * B_1(x_0,\cdot)(\underline{x}) \\ B_2(x_0,\underline{x}) = a_0(\cdot) * B_1(x_0,\cdot)(\underline{x}) = b_0(\cdot) * A_1(x_0,\cdot)(\underline{x}) \end{cases}$$

are conjugate harmonic potentials in $\mathbb{R}^{m+1}_+$ of the function $C_1(x_0,\underline{x})$. It follows that

$$C_2(x_0,\underline{x}) = \frac{1}{2}A_2(x_0,\underline{x}) + \frac{1}{2}\overline{e_0}B_2(x_0,\underline{x})$$

is a monogenic potential in $\mathbb{R}^{m+1}_+$ of C_1.The distributional limits for $x_0 \to 0+$ are given by

$$\begin{cases} a_2(\underline{x}) = \lim\limits_{x_0\to 0+} A_2(x_0,\underline{x}) = a_0 * a_1(\underline{x}) = b_0 * b_1(\underline{x}) \\ b_2(\underline{x}) = \lim\limits_{x_0\to 0+} B_2(x_0,\underline{x}) = a_0 * b_1(\underline{x}) = b_0 * a_1(\underline{x}) \end{cases}$$

which may be calculated explicitly to be

$$a_2(\underline{x}) = -\frac{1}{\pi}\frac{1}{(m-1)(m-3)}\frac{1}{\sigma_{m+1}}T^*_{-m+3}$$

and

$$b_2(\underline{x}) = \frac{1}{2\pi^2}\frac{1}{\sigma_m}\frac{1}{m-2}U^*_{-m+3}.$$

They show the following properties.

Lemma 4

(i) $-\underline{\partial} a_2 = b_1$, $-\underline{\partial} b_2 = a_1$;
(ii) $\mathscr{H}[a_2] = b_2$, $\mathscr{H}[b_2] = a_2$.

The conjugate harmonic potentials $A_2(x_0,\underline{x})$ and $B_2(x_0,\underline{x})$ were also explicitly determined:

$$A_2(x_0,\underline{x}) = \frac{2}{m-1}\frac{1}{\sigma_{m+1}}\frac{x_0}{|\underline{x}|^{m-2}}F_{m-2}\left(\frac{|\underline{x}|}{x_0}\right) - \frac{2}{m-1}\frac{1}{m-3}\frac{1}{\sigma_{m+1}}\frac{1}{|\underline{x}|^{m-3}}$$

$$B_2(x_0,\underline{x}) = \frac{1}{\sigma_{m+1}}\frac{\underline{x}|x|^2}{|\underline{x}|^m}F_m\left(\frac{|\underline{x}|}{x_0}\right) - \frac{m-3}{m-1}\frac{1}{\sigma_{m+1}}\frac{\underline{x}}{|\underline{x}|^{m-2}}F_{m-2}\left(\frac{|\underline{x}|}{x_0}\right).$$

For general $k = 1,2,3,\ldots$, the following functions in $\mathbb{R}^{m+1}_+$ are defined recursively, the convolutions being taken in the variable $\underline{x} \in \mathbb{R}^m$:

$$\begin{aligned} A_k(x_0,\underline{x}) &= a_0 * A_{k-1} = a_1 * A_{k-2} = \cdots = a_{k-1} * A_0 \\ &= b_0 * B_{k-1} = b_1 * B_{k-2} = \cdots = b_{k-1} * B_0 \\ B_k(x_0,\underline{x}) &= a_0 * B_{k-1} = a_1 * B_{k-2} = \cdots = a_{k-1} * B_0 \\ &= b_0 * A_{k-1} = b_1 * A_{k-2} = \cdots = b_{k-1} * A_0 \end{aligned}$$

and

$$C_k(x_0,\underline{x}) = \frac{1}{2}A_k(x_0,\underline{x}) + \frac{1}{2}\overline{e_0}B_k(x_0,\underline{x}).$$

It may be verified that $A_k(x_0,\underline{x})$ and $B_k(x_0,\underline{x})$ are conjugate harmonic potentials of $C_{k-1}(x_0,\underline{x})$ in $\mathbb{R}^{m+1}_+$, while $C_k(x_0,\underline{x})$ is a monogenic potential of $C_{k-1}(x_0,\underline{x})$ in $\mathbb{R}^{m+1}_+$. Their distributional boundary values for $x_0 \to 0+$ are given by the recurrence relations

$$\begin{aligned} a_k(\underline{x}) &= a_0 * a_{k-1} = a_1 * a_{k-2} = \cdots = a_{k-1} * a_0 \\ &= b_0 * b_{k-1} = b_1 * b_{k-2} = \cdots = b_{k-1} * b_0 \\ b_k(\underline{x}) &= a_0 * b_{k-1} = a_1 * b_{k-2} = \cdots = a_{k-1} * b_0 \\ &= b_0 * a_{k-1} = b_1 * a_{k-2} = \cdots = b_{k-1} * a_0 \end{aligned}$$

for which the following explicit formulae may be deduced:

$$\begin{cases} a_{2k} = -\dfrac{1}{2^{2k+1}} \dfrac{\Gamma(\frac{m-2k-1}{2})}{\pi^{\frac{m+2k+1}{2}}} T^*_{-m+2k+1} \\ a_{2k-1} = \dfrac{1}{2^{2k}} \dfrac{\Gamma(\frac{m-2k}{2})}{\pi^{\frac{m+2k}{2}}} T^*_{-m+2k} \end{cases}$$

$$\begin{cases} b_{2k} = \dfrac{1}{2^{2k+1}} \dfrac{\Gamma(\frac{m-2k}{2})}{\pi^{\frac{m+2k+2}{2}}} U^*_{-m+2k+1} \\ b_{2k-1} = -\dfrac{1}{2^{2k}} \dfrac{\Gamma(\frac{m-2k+1}{2})}{\pi^{\frac{m+2k+1}{2}}} U^*_{-m+2k} \end{cases}$$

These distributional limits show the following properties.

Lemma 5 *One has for* $k = 1, 2, \ldots$:

(i) $-\underline{\partial} a_k = b_{k-1}$; $-\underline{\partial} b_k = a_{k-1}$;
(ii) $\mathscr{H}[a_k] = b_{-1} * a_k = b_k$; $\mathscr{H}[b_k] = b_{-1} * b_k = a_k$.

4 Powers of the Dirac Operator

The complex power of the Dirac operator $\underline{\partial}$ was already introduced in [8] and further studied in [6]. It is a convolution operator defined by

$$\begin{aligned} \underline{\partial}^{\mu}[\,.\,] &= \underline{\partial}^{\mu}\delta * [\,.\,] \\ &= \left[\frac{1+e^{i\pi\mu}}{2} \frac{2^{\mu}\Gamma(\frac{m+\mu}{2})}{\pi^{\frac{m-\mu}{2}}} T^*_{-m-\mu} - \frac{1-e^{i\pi\mu}}{2} \frac{2^{\mu}\Gamma(\frac{m+\mu+1}{2})}{\pi^{\frac{m-\mu+1}{2}}} U^*_{-m-\mu} \right] * [\,.\,] \\ &= \frac{2^{\mu}}{\pi^{\frac{m}{2}}} \operatorname{Fp} \frac{1}{|\underline{x}|^{\mu+m}} \left[\frac{1+e^{i\pi\mu}}{2} \frac{\Gamma(\frac{m+\mu}{2})}{\Gamma(-\frac{\mu}{2})} - \frac{1-e^{i\pi\mu}}{2} \frac{\Gamma(\frac{m+\mu+1}{2})}{\Gamma(-\frac{\mu-1}{2})} \underline{\omega} \right] * [\,.\,] \end{aligned} \quad (4)$$

In particular for integer values of the parameter μ, the convolution kernel $\underline{\partial}^{\mu}\delta$ is given by

$$\begin{cases} \underline{\partial}^{2k}\delta = \dfrac{2^{2k}\Gamma(\frac{m+2k}{2})}{\pi^{\frac{m-2k}{2}}} T^*_{-m-2k} \\ \underline{\partial}^{2k+1}\delta = -\dfrac{2^{2k+1}\Gamma(\frac{m+2k+2}{2})}{\pi^{\frac{m-2k}{2}}} U^*_{-m-2k-1}. \end{cases} \tag{5}$$

Note that for $k \in \mathbb{N}_0$ the above expressions (5) are in accordance with the definitions (1) and (2). Moreover, if the dimension m is odd, also all negative integer powers of the Dirac operator are defined by (5). However, if the dimension m is even, the expressions (5) are no longer valid for $k = -\frac{m}{2} - n$, with $n = 0, 1, 2, \ldots$ in the case of $\underline{\partial}^{2k}\delta$ and $n = 1, 2, \ldots$ in the case of $\underline{\partial}^{2k+1}\delta$. Summarizing, $\underline{\partial}^{\mu}$ is defined for all $\mu \in \mathbb{C}$, except for $\mu = -m, -m-1, \ldots$ when m is even. We will define $\underline{\partial}^{\mu}$ for those exceptional values further on. First we prove the following fundamental property.

Proposition 2 *For $\mu, \nu \in \mathbb{C}$ when m is odd or for $\mu, \nu \in \mathbb{C}$ such that μ, ν and $\mu+\nu$ are different from $-m, -m-1, -m-2, \ldots$ when m is even, one has*

$$\underline{\partial}^{\mu}\delta * \underline{\partial}^{\nu}\delta = \underline{\partial}^{\mu+\nu}\delta$$

Proof Using definition (4) for $\underline{\partial}^{\mu}\delta$ and $\underline{\partial}^{\nu}\delta$, the convolution at the left-hand side decomposes into four terms. They are respectively given by

$$\frac{1+e^{i\pi\mu}}{2}\frac{1+e^{i\pi\nu}}{2} 2^{\mu+\nu}\pi^{\frac{m}{2}} \frac{\Gamma(\frac{m+\mu+\nu}{2})}{\pi^{\frac{m-\mu}{2}}\pi^{\frac{m-\nu}{2}}} T^*_{-m-\mu-\nu}$$

for the first one,

$$-\frac{1+e^{i\pi\mu}}{2}\frac{1-e^{i\pi\nu}}{2} 2^{\mu+\nu}\pi^{\frac{m}{2}} \frac{\Gamma(\frac{m+\mu+\nu+1}{2})}{\pi^{\frac{m-\mu}{2}}\pi^{\frac{m-\nu+1}{2}}} U^*_{-m-\mu-\nu}$$

for the second,

$$-\frac{1-e^{i\pi\mu}}{2}\frac{1+e^{i\pi\nu}}{2} 2^{\mu+\nu}\pi^{\frac{m}{2}} \frac{\Gamma(\frac{m+\mu+\nu+1}{2})}{\pi^{\frac{m-\mu+1}{2}}\pi^{\frac{m-\nu}{2}}} U^*_{-m-\mu-\nu}$$

for the third, and

$$\frac{1-e^{i\pi\mu}}{2}\frac{1-e^{i\pi\nu}}{2} 2^{\mu+\nu}\pi^{\frac{m}{2}} \frac{\Gamma(\frac{m+\mu+\nu}{2})}{\pi^{\frac{m-\mu}{2}}\pi^{\frac{m-\nu}{2}}} T^*_{-m-\mu-\nu}$$

for the fourth. The sum of the first and the fourth term thus equals

$$\frac{1+e^{i\pi(\mu+\nu)}}{2} 2^{\mu+\nu} \frac{\Gamma(\frac{m+\mu+\nu}{2})}{\pi^{\frac{m-\mu-\nu}{2}}} T^*_{-m-\mu-\nu}$$

while the sum of the second and the third term equals

$$-\frac{1-e^{i\pi(\mu+\nu)}}{2} 2^{\mu+\nu} \frac{\Gamma(\frac{m+\mu+\nu+1}{2})}{\pi^{\frac{m-\mu-\nu+1}{2}}} U^*_{-m-\mu-\nu}.$$

The sum of the latter two expressions is exactly $\underline{\partial}^{\mu+\nu}\delta$. □

Corollary 1 *For $\mu \in \mathbb{C}$ when m is odd or for $\mu \in \mathbb{C}\setminus\{\pm m, \pm m \pm 1, \pm m \pm 2, \ldots\}$ when m is even, one has*

$$\underline{\partial}^{\mu}\delta * \underline{\partial}^{-\mu}\delta = \delta.$$

Now we put for $\mu \in \mathbb{C}$ when m is odd or for $\mu \in \mathbb{C}\setminus\{m, m+1, m+2, \ldots\}$ when m is even

$$\begin{aligned} E_{\mu} &= \underline{\partial}^{-\mu}\delta \\ &= \frac{1+e^{-i\pi\mu}}{2}\frac{2^{-\mu}\Gamma(\frac{m-\mu}{2})}{\pi^{\frac{m+\mu}{2}}}T^{*}_{-m+\mu} - \frac{1-e^{-i\pi\mu}}{2}\frac{2^{-\mu}\Gamma(\frac{m-\mu+1}{2})}{\pi^{\frac{m+\mu+1}{2}}}U^{*}_{-m+\mu} \end{aligned}$$

and in particular for $k \in \mathbb{Z}$ when m is odd or for $k \in \mathbb{Z}\setminus\{\frac{m}{2}+n, n = 0, 1, 2, \ldots\}$ when m is even

$$\begin{cases} E_{2k} = \dfrac{1}{2^{2k}}\dfrac{\Gamma(\frac{m-2k}{2})}{\pi^{\frac{m+2k}{2}}}T^{*}_{-m+2k} \\ E_{2k+1} = -\dfrac{1}{2^{2k+1}}\dfrac{\Gamma(\frac{m-2k}{2})}{\pi^{\frac{m+2k+2}{2}}}U^{*}_{-m+2k+1}. \end{cases}$$

Then Corollary 1 implies that, for $\mu \in \mathbb{C}$ when m is odd or for $\mu \in \mathbb{C}\setminus\{\pm m, \pm m \pm 1, \pm m \pm 2, \ldots\}$ when m is even, $E_{\mu} = \underline{\partial}^{-\mu}\delta$ is the fundamental solution of the operator $\underline{\partial}^{\mu}$:

$$\underline{\partial}^{\mu}E_{\mu} = \underline{\partial}^{\mu}\delta * E_{\mu} = \delta$$

This is in accordance with a result in [6].

It is also clear that, in the case where the dimension m is even, once the fundamental solutions E_{m+n} of $\underline{\partial}^{m+n}$, $n = 0, 1, 2, \ldots$ are known, we can use these expressions for defining the operators $\underline{\partial}^{-m-n}$, $n = 0, 1, 2, \ldots$. To that end we recall a result of [6].

Proposition 3 *If the dimension m is even, for $n = 0, 1, 2, \ldots$, the fundamental solution E_{m+n} of the operator $\underline{\partial}^{m+n}$ is given by*

$$\begin{cases} E_{m+2j} = (p_{2j}\ln r + q_{2j})T^{*}_{2j} \\ E_{m+2j+1} = (p_{2j+1}\ln r + q_{2j+1})U^{*}_{2j+1} \end{cases} \qquad j = 0, 1, 2, \ldots$$

where the constants p_n and q_n satisfy the recurrence relations

$$\begin{cases} p_{2j+2} = \dfrac{1}{2j+2}p_{2j+1} \\ q_{2j+2} = \dfrac{1}{2j+2}\left(q_{2j+1} - \dfrac{1}{2j+2}p_{2j+1}\right) \end{cases} \qquad j = 0, 1, 2, \ldots$$

and

$$\begin{cases} p_{2j+1} = -\dfrac{1}{2\pi}p_{2j} \\ q_{2j+1} = -\dfrac{1}{2\pi}\left(q_{2j} - \dfrac{1}{m+2j}p_{2j}\right) \end{cases} \qquad j = 0, 1, 2, \ldots$$

with starting values $p_0 = -\frac{1}{2^{m-1}\pi^m}$ and $q_0 = 0$.

Now putting, for m even and $n = 0, 1, 2, \ldots$, $\underline{\partial}^{-m-n}\delta = E_{m+n}$, and hence

$$\underline{\partial}^{-m-n}[\,.\,] = \underline{\partial}^{-m-n}\delta * [\,.\,] = E_{m+n} * [\,.\,]$$

we indeed have

$$\underline{\partial}^{-m-n} E_{-m-n} = \underline{\partial}^{-m-n}\delta * \underline{\partial}^{m+n}\delta = E_{m+n} * \underline{\partial}^{m+n}\delta = \delta.$$

So the operator $\underline{\partial}^{\mu}[\,.\,]$ eventually is defined for all $\mu \in \mathbb{C}$, and there holds in distributional sense

$$\underline{\partial}^{\mu}[E_{\mu}] = \underline{\partial}^{\mu}\left[\underline{\partial}^{-\mu}\delta\right] = \delta, \quad \mu \in \mathbb{C}$$

or, at the level of the operators: $\underline{\partial}^{\mu}\underline{\partial}^{-\mu} = \mathbf{1}$.

5 A New Operator

Recalling the following distributional boundary values of the conjugate harmonic potentials studied in [7]

$$\begin{cases} a_{2k-1} = \dfrac{1}{2^{2k}} \dfrac{\Gamma(\frac{m-2k}{2})}{\pi^{\frac{m+2k}{2}}} T^*_{-m+2k}, & k \in \mathbb{Z},\ 2k < m \\ b_{2k} = \dfrac{1}{2^{2k+1}} \dfrac{\Gamma(\frac{m-2k}{2})}{\pi^{\frac{m+2k+2}{2}}} U^*_{-m+2k+1}, & k \in \mathbb{Z},\ 2k < m \end{cases}$$

it becomes clear, in view of the results in Sect. 4, that these distributional boundary values are nothing but fundamental solutions of appropriate integer powers of the Dirac operator. We have indeed, for integer k such that $2k < m$, that

$$\begin{cases} a_{2k-1} = E_{2k} = \underline{\partial}^{-2k}\delta \\ b_{2k} = -E_{2k+1} = -\underline{\partial}^{-2k-1}\delta \end{cases}$$

showing that at the same time they are also distributions resulting from the action of the opposite integer powers of the Dirac operator on the delta distribution.

It also becomes clear that the other distributional boundary values a_{2k} and b_{2k-1} cannot be expressed in a similar way as fundamental solutions of integer powers of the Dirac operator. Whence the need for a new operator, depending upon a complex parameter μ, the fundamental solutions of which correspond to those distributional boundary values a_{2k} and b_{2k-1}. Taking into account the Hilbert pair relationship between the distributional boundary values, we define to that end the operator ${}^{\mu}\mathscr{H}$ by

$${}^{\mu}\mathscr{H}[\,.\,] = \underline{\partial}^{\mu} H * [\,.\,]$$

where the convolution kernel $\underline{\partial}^{\mu} H$ is given by

$$\underline{\partial}^{\mu} H = \frac{1 - e^{i\pi\mu}}{2} \frac{2^{\mu}\Gamma(\frac{m+\mu}{2})}{\pi^{\frac{m-\mu}{2}}} T^*_{-m-\mu} - \frac{1 + e^{i\pi\mu}}{2} \frac{2^{\mu}\Gamma(\frac{m+\mu+1}{2})}{\pi^{\frac{m-\mu+1}{2}}} U^*_{-m-\mu}.$$

The notation for this new kernel is motivated by the fact that, as shown by a straightforward calculation, it may indeed be obtained as $\underline{\partial}^{\mu} H = \underline{\partial}^{\mu} \delta * H$. In particular for integer values of the parameter μ, the convolution kernel $\underline{\partial}^{\mu} H$ reduces to

$$\begin{cases} \underline{\partial}^{2k} H = -2^{2k} \dfrac{\Gamma(\frac{m+2k+1}{2})}{\pi^{\frac{m-2k+1}{2}}} U^{*}_{-m-2k} \\ \underline{\partial}^{2k+1} H = 2^{2k+1} \dfrac{\Gamma(\frac{m+2k+1}{2})}{\pi^{\frac{m-2k-1}{2}}} T^{*}_{-m-2k-1} \end{cases} \tag{6}$$

with $2k \neq -m-1, -m-3, \ldots$ when m is odd. Note that for $\mu = 0$ the operator ${}^{0}\mathscr{H}$ reduces to the Hilbert transform, while for $\mu = 1$ the so-called Hilbert-Dirac operator (see [2]) is obtained:

$${}^{1}\mathscr{H}[\,.\,] = (-\Delta_m)^{\frac{1}{2}}[\,.\,] = \underline{\partial} H * [\,.\,] = 2\frac{\Gamma(\frac{m+1}{2})}{\pi^{\frac{m-1}{2}}} T^{*}_{-m-1} * [\,.\,]$$

More generally, we also have for integer k such that $2k \neq -m-1, -m-3, \ldots$ when m is odd,

$${}^{2k+1}\mathscr{H}[\,.\,] = \underline{\partial}^{2k+1} H * [\,.\,] = (-\Delta_m)^{k+\frac{1}{2}}[\,.\,]$$

Summarizing, the operator ${}^{\mu}\mathscr{H}$ is defined for all complex values of the parameter μ except for $\mu = -m, -m-1, -m-2, \ldots$ when m is odd. We will use the same method as above, via the fundamental solutions, to define ${}^{\mu}\mathscr{H}$ for those exceptional values.

Proposition 4 *For $\mu, \nu \in \mathbb{C}$ when m is even or for $\mu, \nu \in \mathbb{C}$ such that μ, ν and $\mu + \nu$ are different from $-m, -m-1, -m-2, \ldots$ when m is odd, one has*

$$\underline{\partial}^{\mu} H * \underline{\partial}^{\nu} H = \underline{\partial}^{\mu+\nu} H.$$

Proof The proof is similar to that of Proposition 2. □

Corollary 2 *For $\mu \in \mathbb{C}$ when m is even or for $\mu \in \mathbb{C} \backslash \{\pm m, \pm m \pm 1, \pm m \pm 2, \ldots\}$ when m is odd, one has*

$$\underline{\partial}^{\mu} H * \underline{\partial}^{-\mu} H = \delta.$$

Now we put for $\mu \in \mathbb{C}$ when m is even or for $\mu \in \mathbb{C} \backslash \{m, m+1, m+2, \ldots\}$ when m is odd

$$\begin{aligned} F_{\mu} &= \underline{\partial}^{-\mu} H \\ &= \frac{1 - e^{-i\pi\mu}}{2} \frac{2^{-\mu}\Gamma(\frac{m-\mu}{2})}{\pi^{\frac{m+\mu}{2}}} T^{*}_{-m+\mu} - \frac{1 + e^{-i\pi\mu}}{2} \frac{2^{-\mu}\Gamma(\frac{m-\mu+1}{2})}{\pi^{\frac{m+\mu+1}{2}}} U^{*}_{-m+\mu} \end{aligned}$$

and in particular for integer values of the parameter μ

$$\begin{cases} F_{2k} = -\dfrac{1}{2^{2k}} \dfrac{\Gamma(\frac{m-2k+1}{2})}{\pi^{\frac{m+2k+1}{2}}} U^*_{-m+2k}, \\ k \neq \dfrac{m+1}{2} + n, \ n = 0, 1, \ldots \text{ for } m \text{ odd} \\ F_{2k+1} = \dfrac{1}{2^{2k+1}} \dfrac{\Gamma(\frac{m-2k-1}{2})}{\pi^{\frac{m+2k+1}{2}}} T^*_{-m+2k+1}, \\ k \neq \dfrac{m-1}{2} + n, \ n = 0, 1, \ldots \text{ for } m \text{ odd} \end{cases}$$

Then Corollary 2 implies that for $\mu \in \mathbb{C}$ when m is even or for $\mu \in \mathbb{C} \setminus \{\pm m, \pm m \pm 1, \pm m \pm 2, \ldots\}$ when m is odd

$$^{\mu}\mathcal{H}[F_\mu] = \underline{\partial}^\mu H * F_\mu = \delta$$

expressing that $F_\mu = \underline{\partial}^{-\mu} H$ is the fundamental solution of the operator $\underline{\partial}^\mu \mathcal{H}$ for the allowed values of μ. So it becomes clear that, in the case where the dimension m is odd, if we succeed in establishing the fundamental solutions $F_{m+n}, n = 0, 1, 2, \ldots$ of the corresponding operators $^{m+n}\mathcal{H}$, we can use these expressions for defining the operators $^{-m-n}\mathcal{H}$. We first prove that E_μ and F_μ form a Hilbert pair.

Proposition 5 *For $\mu \in \mathbb{C} \setminus \{m, m+1, m+2, \ldots\}$ one has*

$$\mathcal{H}[E_\mu] = F_\mu$$

Proof For the allowed values of μ we consecutively have

$$\mathcal{H}[E_\mu] = {}^{0}\mathcal{H}[E_\mu] = H * E_\mu = H * \underline{\partial}^{-\mu}\delta = \underline{\partial}^{-\mu}\delta * H = \underline{\partial}^{-\mu} H = F_\mu \qquad \square$$

Now we determine the fundamental solutions F_{m+n}, $n = 0, 1, 2, \ldots$ when the dimension m is odd. The general expression for $\underline{\partial}^{-\mu}$ being no longer valid in that case, this needs a specific approach, which is similar to the one used for determining the fundamental solutions E_{m+n} of $\underline{\partial}^{-m-n}$ when m was even.

Proposition 6 *If the dimension m is odd, then, for $n = 0, 1, 2, \ldots$, the fundamental solution of $^{m+n}\mathcal{H}$ is given by*

$$\begin{cases} F_{m+2j} = (p_{2j} \ln r + q_{2j}) T^*_{2j} \\ F_{m+2j+1} = (p_{2j+1} \ln r + q_{2j+1}) U^*_{2j+1} \end{cases} \qquad j = 0, 1, 2, \ldots$$

with the same constants (p_n, q_n) as in Proposition 3.

Proof We have to prove that $^{m+n}\mathcal{H}[F_{m+n}] = \delta$ or $\underline{\partial}^{m+n} H * F_{m+n} = \delta$, or still $\underline{\partial}^{m+n} * F_{m+n} = H$, which will be satisfied if $\underline{\partial} F_{m+n} = F_{m+n-1}$. If n is even, say $n = 2j$, we have

$$\begin{aligned}\underline{\partial} F_{m+2j} &= p_{2j}\frac{x}{r^2}T^*_{2j} + (p_{2j}\ln r + q_{2j})\underline{\partial} T^*_{2j}\\ &= p_{2j}U^*_{2j-1} + 2j(p_{2j}\ln r + q_{2j})U^*_{2j-1}\end{aligned}$$

from which it follows that the following recurrence relations should hold

$$\begin{cases} p_{2j} + 2jq_{2j} = q_{2j-1} \\ 2jp_{2j} = p_{2j-1} \end{cases} \quad \text{or} \quad \begin{cases} p_{2j} = \dfrac{1}{2j}p_{2j-1} \\ q_{2j} = \dfrac{1}{2j}\left(q_{2j-1} - \dfrac{1}{2j}p_{2j-1}\right) \end{cases}$$

If n is odd, say $n = 2j + 1$, we have

$$\begin{aligned}\underline{\partial} F_{m+2j+1} &= p_{2j+1}\frac{x}{r^2}U^*_{2j+1} + (p_{2j+1}\ln r + q_{2j+1})\underline{\partial} U^*_{2j+1}\\ &= -p_{2j+1}\frac{2\pi}{m+2j}T^*_{2j} - 2\pi(p_{2j+1}\ln r + q_{2j+1})T^*_{2j}\end{aligned}$$

leading to the recurrence relations

$$\begin{cases} -\dfrac{2\pi}{m+2j}p_{2j+1} - 2\pi q_{2j+1} = q_{2j} \\ -2\pi p_{2j+1} = p_{2j} \end{cases} \quad \text{or}$$

$$\begin{cases} p_{2j+1} = -\dfrac{1}{2\pi}p_{2j} \\ q_{2j+1} = -\dfrac{1}{2\pi}\left(q_{2j} - \dfrac{1}{m+2j}p_{2j}\right) \end{cases} \qquad \square$$

So putting for m odd and $n = 0, 1, 2, \ldots$, $\underline{\partial}^{-m-n} H = F_{m+n}$, and hence

$$^{-m-n}\mathscr{H}[\,.\,] = \underline{\partial}^{-m-n} H * [\,.\,] = F_{m+n} * [\,.\,]$$

we indeed have

$$^{-m-n}\mathscr{H}[F_{-m-n}] = \underline{\partial}^{-m-n} H * \underline{\partial}^{m+n} H = \delta$$

Eventually the operator $^{\mu}\mathscr{H}$ is defined for all $\mu \in \mathbb{C}$, and there holds in distributional sense

$$^{\mu}\mathscr{H}[F_\mu] = {}^{\mu}\mathscr{H}\left[{}^{-\mu}H\right] = \delta$$

or, at the level of operators: $^{\mu}\mathscr{H}\,{}^{-\mu}\mathscr{H} = \mathbf{1}$.

Now we expect the distributional boundary values

$$\begin{cases} a_{2k} = -\dfrac{1}{2^{2k+1}}\dfrac{\Gamma(\frac{m-2k-1}{2})}{\pi^{\frac{m+2k+1}{2}}}T^*_{-m+2k+1}, & k \in \mathbb{Z},\ 2k+1 < m \\ b_{2k-1} = -\dfrac{1}{2^{2k}}\dfrac{\Gamma(\frac{m-2k+1}{2})}{\pi^{\frac{m+2k+1}{2}}}U^*_{-m+2k}, & k \in \mathbb{Z},\ 2k-1 < m \end{cases}$$

to be fundamental solutions of $^{\mu}\mathscr{H}$ for specific values of μ. This is indeed the case since

$$\begin{cases} a_{2k} = -F_{2k+1} = -\underline{\partial}^{-2k-1} H, & k \in \mathbb{Z},\ 2k+1 < m \\ b_{2k-1} = F_{2k} = \underline{\partial}^{-2k} H, & k \in \mathbb{Z},\ 2k-1 < m \end{cases}$$

We conclude that all distributional boundary values of the sequence of conjugate harmonic potentials of Sect. 3 are fundamental solutions of $\underline{\partial}^{\mu}$ and ${}^{\mu}\mathscr{H}$ for specific integer values of μ.

6 Powers of the Laplace Operator

For complex powers of the Laplace operator the standard definition (see [11]) reads $(-\Delta_m)^{\beta}[\,.\,] = (-\Delta_m)^{\beta}\delta * [\,.\,]$, where the convolution kernel $(-\Delta_m)^{\beta}\delta$ is given by

$$(-\Delta_m)^{\beta}\delta = 2^{2\beta}\frac{\Gamma(\frac{m+2\beta}{2})}{\pi^{\frac{m-2\beta}{2}}}T^*_{-m-2\beta}.$$

Whence apparently $(-\Delta_m)^{\beta}$ is defined for all complex values of the parameter β, except for $\beta = -\frac{m}{2}, -\frac{m}{2}-1, -\frac{m}{2}-2, \ldots$.

In particular for integer values k of the parameter β, except for $k = -\frac{m}{2}, -\frac{m+2}{2}, -\frac{m+4}{2}, \ldots$ in case the dimension m is even, we have, also in view of (5),

$$(-\Delta_m)^{k}\delta = 2^{2k}\frac{\Gamma(\frac{m+2k}{2})}{\pi^{\frac{m-2k}{2}}}T^*_{-m-2k} = \underline{\partial}^{2k}\delta \tag{7}$$

which is in accordance with the factorization of the Laplace operator by the Dirac operator. In case the dimension m is odd, we have, also in view of (6),

$$(-\Delta_m)^{k+\frac{1}{2}}\delta = 2^{2k+1}\frac{\Gamma(\frac{m+2k+1}{2})}{\pi^{\frac{m-2k-1}{2}}}T^*_{-m-2k-1} = \underline{\partial}^{2k+1}H \tag{8}$$

for integer k, except for $-\frac{m+1}{2}, -\frac{m+3}{2}, -\frac{m+5}{2}, \ldots$.

By a straightforward calculation, similar to the one in the proof of Proposition 2, the following fundamental property is proven.

Proposition 7 *For $\alpha, \beta \in \mathbb{C}$ such that α, β and $\alpha+\beta$ are different from $-\frac{m}{2}-n$, $n = 0, 1, 2, \ldots$ one has*

$$(-\Delta_m)^{\alpha}\delta * (-\Delta_m)^{\beta}\delta = (-\Delta_m)^{\alpha+\beta}\delta$$

Corollary 3 *For $\beta \in \mathbb{C}\setminus\{\pm\frac{m}{2}, \pm\frac{m+2}{2}, \pm\frac{m+4}{2}, \ldots\}$ one has*

$$(-\Delta_m)^{\beta}\delta * (-\Delta_m)^{-\beta}\delta = \delta$$

Now putting for $\beta \in \mathbb{C}\setminus\{\frac{m}{2}, \frac{m+2}{2}, \frac{m+4}{2}, \ldots\}$

$$K_{\beta} = (-\Delta_m)^{-\beta}\delta = 2^{-\beta}\frac{\Gamma(\frac{m-2\beta}{2})}{\pi^{\frac{m+2\beta}{2}}}T^*_{-m+2\beta}$$

and in particular for integer k

$$\begin{cases} K_k = (-\Delta_m)^{-k}\delta = \dfrac{1}{2^{2k}} \dfrac{\Gamma(\frac{m-2k}{2})}{\pi^{\frac{m+2k}{2}}} T^*_{-m+2k}, & 2k < m \text{ when } m \text{ is even} \\ K_{k+\frac{1}{2}} = (-\Delta_m)^{-k-\frac{1}{2}}\delta = \dfrac{1}{2^{2k+1}} \dfrac{\Gamma(\frac{m-2k-1}{2})}{\pi^{\frac{m+2k+1}{2}}} T^*_{-m+2k+1}, & \\ & 2k < m-1 \text{ when } m \text{ is odd} \end{cases} \tag{9}$$

the above Corollary 3 implies that

$$(-\Delta_m)^{\beta}[K_\beta] = \delta, \quad \beta \in \mathbb{C}\backslash\left\{\pm\frac{m}{2}, \pm\frac{m+2}{2}, \pm\frac{m+4}{2}, \ldots\right\}$$

which expresses the fact that $K_\beta = (-\Delta_m)^{-\beta}\delta$ is the fundamental solution of the operator $(-\Delta_m)^\beta$ for $\beta \in \mathbb{C}\backslash\{\pm\frac{m}{2}, \pm\frac{m+2}{2}, \pm\frac{m+4}{2}, \ldots\}$.

We still need to define the operator $(-\Delta_m)^\beta$ for $\beta = -\frac{m}{2}, -\frac{m+2}{2}, -\frac{m+4}{2}, \ldots$ and the fundamental solution K_β for $\beta = \frac{m}{2}, \frac{m+2}{2}, \frac{m+4}{2}, \ldots$. Keeping in mind the formulae (7) and (8), which we still want to remain valid, we put, for $n = 0, 1, 2, \ldots$

(i) when m is odd:

$$(-\Delta_m)^{-\frac{m}{2}-n}\delta = \underline{\partial}^{-m-2n}H = F_{m+2n} = (p_{2n}\ln r + q_{2n})T^*_{2n}$$

and $K_{\frac{m}{2}+n} = F_{m+2n}$;

(ii) when m is even:

$$(-\Delta_m)^{-\frac{m}{2}-n}\delta = \underline{\partial}^{-m-2n}\delta = E_{m+2n} = (p_{2n}\ln r + q_{2n})T^*_{2n}$$

and $K_{\frac{m}{2}+n} = E_{m+2n}$.

We then indeed have

(i) for m odd

$$(-\Delta_m)^{\frac{m}{2}+n}[K_{\frac{m}{2}+n}] = \underline{\partial}^{m+2n}H * K_{\frac{m}{2}+n} = \underline{\partial}^{m+2n}H * F_{m+2n} = \delta$$

(ii) for m even

$$(-\Delta_m)^{\frac{m}{2}+n}[K_{\frac{m}{2}+n}] = \underline{\partial}^{m+2n}\delta * K_{\frac{m}{2}+n} = \underline{\partial}^{m+2n}\delta * E_{m+2n} = \delta$$

which eventually leads to

$$(-\Delta_m)^{\frac{m}{2}+n}(-\Delta_m)^{-\frac{m}{2}-n} = \mathbf{1}.$$

Note that for *natural* powers of the Laplace operator, the above fundamental solutions are in accordance with the results of [1], where also the closed form of the coefficients p_{2n} and q_{2n}, $n = 0, 1, 2, \ldots$ can be found.

7 A Second New Operator

The conclusion of Sects. 4 and 5 was that all distributional boundary values of the sequence of conjugate harmonic potentials studied in [7], and recalled in Sect. 3, are fundamental solutions of the operators $\underline{\partial}^{\mu}$ and ${}^{\mu}\mathscr{H}$ for specific integer values of the parameter μ. Wondering if they are also fundamental solutions of the operator $(-\Delta_m)^{\beta}$ for some specific values of the complex parameter β, we indeed find that, in view of (9),

$$\begin{cases} a_{2k} = -\dfrac{1}{2^{2k+1}} \dfrac{\Gamma(\frac{m-2k-1}{2})}{\pi^{\frac{m+2k+1}{2}}} T^*_{-m+2k+1} = -K_{k+\frac{1}{2}} \\ \quad = -(-\Delta_m)^{-k-\frac{1}{2}}\delta, & 2k+1<m \\ a_{2k-1} = \dfrac{1}{2^{2k}} \dfrac{\Gamma(\frac{m-2k}{2})}{\pi^{\frac{m+2k}{2}}} U^*_{-m+2k} = K_k = (-\Delta_m)^{-k}\delta, & 2k<m. \end{cases}$$

To recover the distributional boundary values b_{2k} and b_{2k-1} as fundamental solutions of powers of the Laplace operator, apparently a new operator has to come into play again. Bearing in mind that the distributional boundary values are forming Hilbert pairs, we define the operator ${}^{\beta}\mathscr{L}$ by

$$ {}^{\beta}\mathscr{L}[\,.\,] = (-\Delta_m)^{\beta} H * [\,.\,] $$

where the convolution kernel $(-\Delta_m)^{\beta} H$ is given by

$$ (-\Delta_m)^{\beta} H = -2^{\beta} \frac{\Gamma(\frac{m+2\beta+1}{2})}{\pi^{\frac{m-2\beta+1}{2}}} U^*_{-m-2\beta}. $$

The notation for this second new kernel is motivated by the fact that, as shown by a straightforward calculation, it may indeed be obtained as the convolution $(-\Delta_m)^{\beta} H = (-\Delta_m)^{\beta}\delta * H$. Apparently the operator ${}^{\beta}\mathscr{L}$ is defined for all complex values of the parameter β except for $\beta = -\frac{m+1}{2} - n$, $n = 0, 1, 2, \ldots$.

Note the particular cases for integer k:

$$\begin{cases} (-\Delta_m)^{k} H = -2^{2k} \dfrac{\Gamma(\frac{m+2k+1}{2})}{\pi^{\frac{m-2k+1}{2}}} U^*_{-m-2k}, \\ k \neq -\dfrac{m+1}{2}, -\dfrac{m+3}{2}, \ldots \ (m \text{ odd}) \\ (-\Delta_m)^{k+\frac{1}{2}} H = -2^{2k+1} \dfrac{\Gamma(\frac{m+2k+2}{2})}{\pi^{\frac{m-2k}{2}}} U^*_{-m-2k-1}, \\ k \neq -\dfrac{m+2}{2}, -\dfrac{m+4}{2}, \ldots \ (m \text{ even}). \end{cases}$$

It follows that

$$\begin{cases} (-\Delta_m)^{k} H = \underline{\partial}^{2k} H, & k \in \mathbb{Z},\ k \neq -\dfrac{m+1}{2}, -\dfrac{m+3}{2}, \ldots \ (m \text{ odd}) \\ (-\Delta_m)^{k+\frac{1}{2}} H = \underline{\partial}^{2k+1}\delta, & k \in \mathbb{Z},\ k \neq -\dfrac{m+2}{2}, -\dfrac{m+4}{2}, \ldots \ (m \text{ even}) \end{cases} \tag{10}$$

or, at the level of the operators: ${}^{k}\mathscr{L} = {}^{2k}\mathscr{H}$ and ${}^{k+\frac{1}{2}}\mathscr{L} = \underline{\partial}^{2k+1}$.

Proposition 8 *For $\alpha, \beta \in \mathbb{C}$ such that α, β and $\alpha+\beta$ are different from $-\frac{m+1}{2}-n$, $n=0,1,2,\ldots$, one has*

$$(-\Delta_m)^\alpha H * (-\Delta_m)^\beta H = (-\Delta_m)^{\alpha+\beta}\delta.$$

Proof The proof is similar to the one of Proposition 2. □

Corollary 4 *For $\beta \in \mathbb{C}\backslash\{\pm\frac{m+1}{2} \pm n,\ n=0,1,2,\ldots\}$ one has $(-\Delta_m)^\beta H * (-\Delta_m)^{-\beta} H = \delta$.*

Putting, for $\beta \in \mathbb{C}\backslash\{\frac{m+1}{2}+n,\ n=0,1,2,\ldots\}$,

$$L_\beta = (-\Delta_m)^{-\beta} H = -2^{-\beta}\frac{\Gamma(\frac{m-2\beta+1}{2})}{\pi^{\frac{m+2\beta+1}{2}}}U^*_{-m+2\beta}$$

and in particular for integer k

$$\begin{cases} L_k = -2^{-2k}\dfrac{\Gamma(\frac{m-2k+1}{2})}{\pi^{\frac{m+2k+1}{2}}}U^*_{-m+2k}, & 2k < m+1 \\[2ex] L_{k+\frac{1}{2}} = -2^{-2k+1}\dfrac{\Gamma(\frac{m-2k}{2})}{\pi^{\frac{m+2k+2}{2}}}U^*_{-m+2k+1}, & 2k < m. \end{cases}$$

Corollary 4 implies that

$${}^\beta\mathscr{L}[L_\beta] = \delta, \quad \beta \in \mathbb{C}\Big\backslash\left\{\pm\frac{m+1}{2} \pm n,\ n=0,1,2,\ldots\right\}$$

expressing the fact that $L_\beta = (-\Delta_m)^{-\beta} H$ is the fundamental solution of the operator ${}^\beta\mathscr{L}$ for $\beta \in \mathbb{C}\backslash\{\pm\frac{m+1}{2} \pm n,\ n=0,1,2,,\ldots\}$.

Proposition 9 *For $\beta \in \mathbb{C}\backslash\{\frac{m+n}{2},\ n=0,1,2,\ldots\}$ one has $\mathscr{H}[K_\beta] = L_\beta$.*

Proof For the allowed values of β we consecutively have

$$\mathscr{H}[K_\beta] = {}^0\mathscr{H}[K_\beta] = H * K_\beta = H * (-\Delta_m)^{-\beta}\delta = (-\Delta_m)^{-\beta} H = L_\beta$$ □

Now we define the operator ${}^\beta\mathscr{L}$ for $\beta = -\frac{m+1}{2} - n$, $n=0,1,2,\ldots$ and the fundamental solution L_β for $\beta = \frac{m+1}{2}+n$, $n=0,1,2,\ldots$:

(i) if m is odd, we put

$$(-\Delta_m)^{-\frac{m+1}{2}-n} H = \underline{\partial}^{-m-2n-1} H = F_{m+2n+1} = (p_{2n+1}\ln r + q_{2n+1})U^*_{2n+1}$$

and $L_{\frac{m+1}{2}+n} = F_{m+2n+1}$;

(ii) if m is even, we put

$$(-\Delta_m)^{-\frac{m+1}{2}-n} H = \underline{\partial}^{-m-2n-1}\delta = E_{m+2n+1} = (p_{2n+1}\ln r + q_{2n+1})U^*_{2n+1}$$

and $L_{\frac{m+1}{2}+n} = E_{m+2n+1}$.

In this way the properties (10) are preserved for the exceptional values of β, and moreover

$$^{\frac{m+1}{2}+n}\mathscr{L}[L_{\frac{m+1}{2}+n}] = {}^{\frac{m+1}{2}+n}\mathscr{L}\big[(-\Delta_m)^{-\frac{m+1}{2}-n}H\big] = \delta$$

or

$$\big({}^{\frac{m+1}{2}+n}\mathscr{L}\big)\big({}^{-\frac{m+1}{2}-n}\mathscr{L}\big) = \mathbf{1}.$$

As expected the distributional boundary values b_ℓ are indeed recovered from the fundamental solutions of the operator ${}^{\beta}\mathscr{L}$, since

$$\begin{cases} b_{2k} = -E_{2k+1} = -L_{k+\frac{1}{2}}, & 2k < m \\ b_{2k-1} = F_{2k} = L_k, & 2k < m+1. \end{cases}$$

8 Conclusion

In this paper we have shown that the distributional boundary values for $x_0 \to 0+$ of the sequence of conjugate harmonic potentials in upper half-space $\mathbb{R}^{m+1}_+ = \{x_0e_0 + \underline{x} : \underline{x} \in \mathbb{R}^m,\ x_0 > 0\}$, can be expressed as fundamental solutions of specific powers of four operators: the standard operators $\underline{\partial}^{\mu}$ and $(-\Delta_m)^{\beta}$ and the two newly introduced operators ${}^{\mu}\mathscr{H}$ and ${}^{\beta}\mathscr{L}$. The extension of the definition of those four operators to the exceptional values of the complex parameters μ and β for which the operators were not defined initially, was crucial for this purpose. The unifying character of the families of Clifford distributions $\mathscr{T}^*$ and $\mathscr{U}^*$ in this is remarkable.

For specific values of the complex parameters μ and β, the four operators studied are interconnected. Since these relationships are fundamental we recall them here. For integer k one has for the corresponding convolution kernels:

$$(-\Delta_m)^k\delta = \underline{\partial}^{2k}\delta; \qquad (-\Delta_m)^k H = \underline{\partial}^{2k}H;$$
$$(-\Delta_m)^{k+\frac{1}{2}}\delta = \underline{\partial}^{2k+1}H; \qquad (-\Delta_m)^{k+\frac{1}{2}}H = \underline{\partial}^{2k+1}\delta.$$

The apparent symmetries in these formulae strengthen the idea that, like the Dirac or delta-distribution δ, also the Hilbert kernel H really is a fundamental distribution, more or less a counterpart to the pointly supported δ.

The above formulae also generalize the well-known fact that the composition of the two Clifford vector operators $\underline{\partial}$ and $\mathscr{H}$ equals the scalar operator *square root of the Laplacian* $(-\Delta_m)^{\frac{1}{2}}$:

$$(-\Delta_m)^{\frac{1}{2}}\delta = \underline{\partial}H.$$

This also leads to the well-known scalar factorization of the Laplace operator in terms of pseudodifferential operators: $-\Delta_m = (-\Delta_m)^{\frac{1}{2}}(-\Delta_m)^{\frac{1}{2}}$, next to its vector factorization, used by P.A.M. Dirac under matrix disguise: $-\Delta_m = \underline{\partial}\,\underline{\partial}$. The fact that the Laplace operator may be factorized in two completely different ways is explained by the fact that both the convolution of two T^*-distributions—to which family $(-\Delta_m)^{\frac{1}{2}}$ belongs—and the convolution of two U^*-distributions—to which family $\underline{\partial}$ belongs—result into a T^*-distribution. On the contrary, vector valued operators, such as $\underline{\partial}$ and $\mathscr{H}$, only have one factorization based on the convolution of a T^*-distribution with a U^*-distribution, as shown in the following remarkable formulae:

$$\underline{\partial}\delta = (-\Delta_m)^{\frac{1}{2}} H \qquad H = (-\Delta_m)^{\frac{1}{2}} \delta.$$

Finally note that each of the fundamental solutions of the four operators studied in this paper, or, in other words, each of the distributional boundary values of the conjugate harmonic potentials studied in [7], may be used as a convolution kernel to define an operator of the same kind but with opposite parameter value:

$$\begin{aligned}
\underline{\partial}^{\mu} E_{\mu} = \delta \quad &\text{and} \quad E_{\mu} * [\,.\,] = \underline{\partial}^{-\mu}[\,.\,] \\
\underline{\partial}^{\mu} H F_{\mu} = \delta \quad &\text{and} \quad F_{\mu} * [\,.\,] = {}^{-\mu}\mathscr{H}[\,.\,] \\
(-\Delta_m)^{\beta} K_{\beta} = \delta \quad &\text{and} \quad K_{\beta} * [\,.\,] = (-\Delta_m)^{-\beta}[\,.\,] \\
(-\Delta_m)^{\beta} H L_{\beta} = \delta \quad &\text{and} \quad L_{\beta} * [\,.\,] = {}^{-\beta}\mathscr{L}[\,.\,]
\end{aligned}$$

References

1. Aronszajn, N., Creese, T.M., Lipkin, L.J.: Polyharmonic Functions. Clarendon, Oxford (1983)
2. Brackx, F., De Schepper, H.: Hilbert-Dirac operators in Clifford analysis. Chin. Ann. Math. **26B**(1), 1–14 (2005)
3. Brackx, F., Delanghe, R., Sommen, F.: Clifford Analysis. Pitman, Boston (1982)
4. Brackx, F., Delanghe, R., Sommen, F.: Spherical means, distributions and convolution operators in Clifford analysis. Chin. Ann. Math. **24B**(2), 133–146 (2003)
5. Brackx, F., Delanghe, R., Sommen, F.: Spherical means and distributions in Clifford analysis. In: Qian, T., Hempfling, T., McIntosh, A., Sommen, F. (eds.) Advances in Analysis and Geometry: New Developments Using Clifford Algebra. Trends in Mathematics. Birkhäuser, Basel (2004)
6. Brackx, F., De Knock, B., De Schepper, H., Eelbode, D.: A calculus scheme for Clifford distributions. Tokyo J. Math. **29**(2), 495–513 (2006)
7. Brackx, F., De Bie, H., De Schepper, H.: On a chain of harmonic and monogenic potentials in Euclidean half-space (submitted for publication)
8. Delanghe, R., Sommen, F., Souček, V.: Clifford Algebra and Spinor-Valued Functions—A Function Theory for the Dirac Operator. Kluwer Academic, Dordrecht (1992)
9. Gilbert, J., Murray, M.: Clifford Algebra and Dirac Operators in Harmonic Analysis. Cambridge University Press, Cambridge (1991)
10. Gradshteyn, I.S., Ryzhik, I.M.: Table of Integrals, Series, and Products, 7th edn. Translated from Russian, translation edited and with a preface by A. Jeffrey and D. Zwillinger. Elsevier, Amsterdam (2007)
11. Helgason, S.: Groups and Geometric Analysis. Pure and Applied Mathematics. Academic Press, Orlando (1984)
12. Lang, S.: Complex Analysis. Graduate Texts in Mathematics, vol. 103. Springer, New York (1999)
13. Porteous, I.: Topological Geometry. Van Nostrand-Reinhold, London (1969)

On Two Approaches to the Bergman Theory for Slice Regular Functions

Fabrizio Colombo, J. Oscar González-Cervantes, Maria Elena Luna-Elizarrarás, Irene Sabadini, and Michael Shapiro

Abstract In this paper we show that the classical Bergman theory admits two possible settings for the class of slice regular functions. Let Ω be a suitable open subset of the space of quaternions $\mathbb{H}$ that intersects the real line and let $\mathbb{S}^2$ be the unit sphere of purely imaginary quaternions. Slice regular functions are those functions $f : \Omega \to \mathbb{H}$ whose restriction to the complex planes $\mathbb{C}(\mathbf{i})$, for every $\mathbf{i} \in \mathbb{S}^2$, are holomorphic maps. One of their crucial properties is that from the knowledge of the values of f on $\Omega \cap \mathbb{C}(\mathbf{i})$ for some $\mathbf{i} \in \mathbb{S}^2$, one can reconstruct f on the whole Ω by the so called Representation Formula. We will define the so-called slice regular Bergman theory of the first kind. By the Riesz representation theorem we provide a Bergman kernel which is defined on Ω and is a reproducing kernel. In the slice regular Bergman theory of the second kind we use the Representation Formula to define another Bergman kernel; this time the kernel is still defined on Ω but the integral representation of f requires the calculation of the integral only on $\Omega \cap \mathbb{C}(\mathbf{i})$ and the integral does not depend on $\mathbf{i} \in \mathbb{S}^2$.

M.E. Luna-Elizarrarás and M. Shapiro were partially supported by CONACYT projects as well as by Instituto Politécnico Nacional in the framework of COFAA and SIP programs.

F. Colombo (✉) · I. Sabadini
Department of Mathematics, Politecnico di Milano, Via E. Bonardi 9, 20133 Milan, Italy
e-mail: fabrizio.colombo@polimi.it

I. Sabadini
e-mail: irene.sabadini@polimi.it

J.O. González-Cervantes · M.E. Luna-Elizarrarás · M. Shapiro
Department of Mathematics, E.S.F.M. del I.P.N., Av. Instituto Politécnico Nacional s/n, 07338 Mexico D.F., Mexico

J.O. González-Cervantes
e-mail: jogc200678@gmail.com

M.E. Luna-Elizarrarás
e-mail: eluna@esfm.ipn.mx

M. Shapiro
e-mail: shapiro@esfm.ipn.mx

G. Gentili et al. (eds.), *Advances in Hypercomplex Analysis*, Springer INdAM Series 1,
DOI 10.1007/978-88-470-2445-8_3, © Springer-Verlag Italia 2013

1 Introduction

Bergman theory is an important topic which has been studied in several settings. The literature is very wide and as classical reference books, in the case of holomorphic functions, we mention [1, 2, 19]. The theory has also been developed for hyperholomorphic functions such as the quaternionic regular functions in the sense of Fueter and the theory of monogenic functions. Without claiming completeness we mention [3, 4, 13–16, 18, 20–22] and the literature therein. In this paper we introduce the first elements and the settings of the Bergman theory in the case of slice regular functions studied in the papers [5–7, 9, 17].

More in general, by slice hyperholomorphic functions we mean both slice regular functions and slice monogenic functions which are the monogenic version of the theory of slice regular functions, see [6, 8, 10, 11].

Let us come to some preliminary considerations on the possible definitions of Bergman space in the context of slice regular functions.

Roughly speaking, slice regular functions $f : \Omega \to \mathbb{H}$, where Ω is an open set, are those real differentiable functions whose restrictions to $\Omega_{\mathbf{i}} := \Omega \cap \mathbb{C}(\mathbf{i})$, for every $\mathbf{i} \in \mathbb{S}^2$, are holomorphic maps. In the sequel by $\mathcal{SR}(\Omega)$ we denote the set of slice regular functions on Ω. Let Ω be an open set in $\mathbb{H}$ which is invariant with respect to rotations that fix the real axis, i.e., it is axially symmetric. When we suppose that the domain Ω intersects the real line then the holomorphic functions on two different planes, $\mathbb{C}(\mathbf{i})$ and $\mathbb{C}(\mathbf{j})$ with $\mathbf{i} \neq \mathbf{j}$, turn out to be strictly related by the Representation Formula

$$f(x + y\mathbf{i}) = \frac{1}{2}(1 - \mathbf{ij})f(x + y\mathbf{j}) + \frac{1}{2}(1 + \mathbf{ij})f(x - y\mathbf{j})$$

which asserts that if we know the functions on a complex plane $\mathbb{C}(\mathbf{j})$ then we can reconstruct f in all the other planes $\mathbb{C}(\mathbf{i})$ for every $\mathbf{i} \in \mathbb{S}^2$. Note that it is not restrictive to consider axially symmetric open sets: if the open set Ω which is the domain of a function f is not axially symmetric we can always extend f to the smallest axially symmetric set Ω' containing Ω.

As a consequence of this fact, the Cauchy formula for this class of functions allows to reconstruct a function f from its boundary values chosen on a plane $\mathbb{C}(\mathbf{i})$ where $\mathbf{i} \in \mathbb{S}^2$ is arbitrarily chosen.

Thus the global behavior of these functions on axially symmetric sets is in fact completely determined by their behavior on $\Omega \cap \mathbb{C}(\mathbf{i})$.

We define the *slice regular Bergman space of the first kind* as

$$\mathcal{A}(\Omega) := \left\{ f \in \mathcal{SR}(\Omega) \;\middle|\; \|f\|^2_{\mathcal{A}(\Omega)} := \int_\Omega |f|^2 \, d\mu < \infty \right\}.$$

This gives us the slice regular Bergman kernel of the first kind leading to the integral representation

$$f(q) = \int_\Omega \mathcal{B}_\Omega(q, \cdot) f \, d\mu, \quad \forall f \in \mathcal{A}(\Omega). \tag{1}$$

The function $\mathcal{B}_\Omega(\cdot,\cdot) : \Omega \times \Omega \to \mathbb{H}$ will be called the slice regular Bergman kernel of the first kind associated to Ω.

The second way to define the Bergman space and the Bergman kernel is to work on a complex plane $\mathbb{C}(\mathbf{i})$ and then to extend the kernel by the Representation Formula. Thus we introduce the so called Bergman theory of the second kind in which we use a kernel $\mathscr{K}_\Omega$ that can be written in a closed form for various open sets Ω. The kernel $\mathscr{K}_\Omega$ is a reproducing kernel on each slice $\mathbb{C}(\mathbf{i})$, i.e.,

$$f(q) = \int_{\Omega\cap\mathbb{C}(\mathbf{i})} \mathscr{K}_\Omega(q,\cdot) f \, d\sigma_{\mathbf{i}}$$

and the integral does not depend on $\mathbf{i} \in \mathbb{S}^2$.

Thus there are two mutually complementary theories.

2 Preliminary Results on Slice Regular Functions

In this section we recall some preliminary results on slice regular functions. Set $\mathbb{B}^4 := \{q \in \mathbb{H} \mid |q| < 1\}$, where $|q|$ is the modulus of q, and writing a quaternion as $q = q_0 + \mathbf{q}$ with obvious meanings of the symbols, and set

$$\mathbb{S}^2 := \{\mathbf{q} \in \mathbb{R}^3 \mid |\mathbf{q}|_{\mathbb{R}^3} = 1\}.$$

Given $\mathbf{i} \in \mathbb{S}^2$, by $\mathbb{C}(\mathbf{i})$ we mean the real-linear space generated by 1 and $\mathbf{i}$. Note that $\mathbb{C}(\mathbf{i}) \cong \mathbb{C}$. Let $\Omega \subset \mathbb{H}$ be an open set. Denote

$$\Omega_{\mathbf{i}} := \Omega \cap \mathbb{C}(\mathbf{i}).$$

In particular, $D_{\mathbf{i}} := \mathbb{B}^4 \cap \mathbb{C}(\mathbf{i})$ and

$$D_{\mathbf{i}}^+ := \{x + y\mathbf{i} \mid y \geq 0 \text{ and } x^2 + y^2 < 1\}.$$

Finally, when we consider a function $f : \Omega \to \mathbb{H}$ then we will denote its restriction to $\Omega \cap \mathbb{C}(\mathbf{i})$ in local coordinates by $f_{|\Omega_{\mathbf{i}}}$.

We are now ready to recall the definition of slice regular functions.

Definition 1 A real differentiable quaternion-valued function f defined on an open set $\Omega \subset \mathbb{H}$ is called (left) slice regular on Ω if for any $\mathbf{i} \in \mathbb{S}^2$ the function $f_{|\Omega_{\mathbf{i}}}$ is such that

$$\left(\frac{\partial}{\partial x} + \mathbf{i}\frac{\partial}{\partial y}\right) f_{|\Omega_{\mathbf{i}}}(x + y\mathbf{i}) = 0 \quad \text{on } \Omega_{\mathbf{i}}.$$

The function f is called slice anti-regular on the right if for any $\mathbf{i} \in \mathbb{S}^2$ the function $f_{|\Omega_{\mathbf{i}}}$ is such that

$$f_{|\Omega_{\mathbf{i}}}(x + y\mathbf{i})\left(\frac{\partial}{\partial x} - \mathbf{i}\frac{\partial}{\partial y}\right) = 0 \quad \text{on } \Omega_{\mathbf{i}}.$$

We denote by $\mathscr{SR}(\Omega)$ the set of slice regular functions on Ω.

Remark 1 It is also possible to define slice right regular functions and slice left anti-regular functions, but we will not use them so we omit their definitions.

The domains on which we will consider slice regular functions are described in the definitions below.

Definition 2 Let $\Omega \subseteq \mathbb{H}$ be a domain. We say that Ω is a slice domain (s-domain for short) if $\Omega \cap \mathbb{R}$ is non empty and if $\Omega \cap \mathbb{C}(\mathbf{i})$ is a domain in $\mathbb{C}(\mathbf{i})$ for all $\mathbf{i} \in \mathbb{S}^2$.

Definition 3 Let $\Omega \subseteq \mathbb{H}$. We say that Ω is axially symmetric if for every $x + \mathbf{i}y \in \Omega$ we have that $x + \mathbf{j}y \in \Omega$ for all $\mathbf{j} \in \mathbb{S}^2$.

Any axially symmetric open set Ω can be uniquely associated with an open set $O_\Omega \subseteq \mathbb{R}^2$ defined by

$$O_\Omega := \{(x, y) \in \mathbb{R}^2 \mid x + y\mathbf{i} \in \Omega,\ \mathbf{i} \in \mathbb{S}^2\}.$$

Slice regular functions obey the following property (see [5, 9]).

Theorem 1 (Representation Formula) *Let f be a slice regular function on an axially symmetric s-domain $\Omega \subseteq \mathbb{H}$. Choose any $\mathbf{j} \in \mathbb{S}^2$. Then the following equality holds for all $q = x + \mathbf{i}y \in \Omega$:*

$$f(x + y\mathbf{i}) = \frac{1}{2}(1 - \mathbf{i}\mathbf{j}) f(x + y\mathbf{j}) + \frac{1}{2}(1 + \mathbf{i}\mathbf{j}) f(x - y\mathbf{j}). \tag{2}$$

Remark 2 Since $\mathbf{i}$ varies in $\mathbb{S}^2$ it is not restrictive to assume $y \geq 0$ in formula (2).

3 The Slice Regular Bergman Theory of the First Kind

In the sequel we will consider the right linear space $\mathcal{L}_2(\Omega, \mathbb{H})$ of square integrable functions $f : \Omega \to \mathbb{H}$ equipped with the norm

$$\|f\|_{\mathcal{L}_2(\Omega,\mathbb{H})} := \left(\int_\Omega |f|^2 \, d\mu\right)^{\frac{1}{2}},$$

where $d\mu$ denotes the volume element in $\mathbb{R}^4$. We have the following well known properties: let $f, g \in \mathcal{L}_2(\Omega, \mathbb{H})$ and let $\lambda \in \mathbb{H}$, then

(i) $\|f\lambda\|_{\mathcal{L}_2(\Omega,\mathbb{H})} = \|f\|_{\mathcal{L}_2(\Omega,\mathbb{H})} |\lambda|$.
(ii) $\|f + g\|_{\mathcal{L}_2(\Omega,\mathbb{H})} \leq \|f\|_{\mathcal{L}_2(\Omega,\mathbb{H})} + \|g\|_{\mathcal{L}_2(\Omega,\mathbb{H})}$.
(iii) $\|f\|_{\mathcal{L}_2(\Omega,\mathbb{H})} = 0 \iff f = 0$ a.e. on Ω.

Moreover, it is known that the function $\langle \cdot, \cdot \rangle_{\mathcal{L}_2(\Omega,\mathbb{H})} : \mathcal{L}_2(\Omega, \mathbb{H}) \times \mathcal{L}_2(\Omega, \mathbb{H}) \to \mathbb{H}$, defined by

$$\langle f, g \rangle_{\mathcal{L}_2(\Omega,\mathbb{H})} = \int_\Omega \bar{f} g \, d\mu, \tag{3}$$

is an $\mathbb{H}$-valued inner product on $\mathcal{L}_2(\Omega, \mathbb{H})$. Among its properties we have the following.

Let $f, g, h \in \mathscr{L}_2(\Omega, \mathbb{H})$ and let $q \in \mathbb{H}$ then

(i) $\langle f, g + hq\rangle_{\mathscr{L}_2(\Omega,\mathbb{H})} = \langle f, g\rangle_{\mathscr{L}_2(\Omega,\mathbb{H})} + \langle f, h\rangle_{\mathscr{L}_2(\Omega,\mathbb{H})} q$.
(ii) $\langle f, g\rangle_{\mathscr{L}_2(\Omega,\mathbb{H})} = \overline{\langle g, f\rangle}_{\mathscr{L}_2(\Omega,\mathbb{H})}$.
(iii) $\langle f, f\rangle_{\mathscr{L}_2(\Omega,\mathbb{H})} = \|f\|^2_{\mathscr{L}_2(\Omega,\mathbb{H})} \geq 0$.

Definition 4 The set $\mathscr{A}(\Omega) := \mathscr{SR}(\Omega) \cap \mathscr{L}_2(\Omega, \mathbb{H})$, equipped with the norm and the inner product inherited from $\mathscr{L}_2(\Omega, \mathbb{H})$, is called the slice regular Bergman space of the first kind associated to Ω.

Remark 3 It is immediate to verify that the space $\mathscr{A}(\Omega)$ is a quaternionic right-linear space.

Proposition 1 (Evaluation functional for $\Omega_{\mathbf{i}}$) *Let Ω be a bounded axially symmetric s-domain, let $\mathbf{i} \in \mathbb{S}^2$ and let $K \subset \Omega_{\mathbf{i}}$ be a compact set. Then there exists a constant $\lambda_K > 0$ such that*

$$\sup\{|f(q)| \mid q \in K\} \leq \lambda_K \|f\|_{\mathscr{A}(\Omega)}, \quad \forall f \in \mathscr{A}(\Omega).$$

Proof Using Lemma 1.4.1 given in [19], one can find a constant $\alpha_K > 0$ depending of K, such that

$$\sup\{|g(z)| \mid z \in K\} \leq \alpha_K \left[\int_{\Omega_{\mathbf{i}}} |g|^2 \, d\sigma_{\mathbf{i}}\right]^{\frac{1}{2}},$$

for any function g belonging to the usual holomorphic Bergman space for $\Omega_{\mathbf{i}}$.

Since every $f \in \mathscr{A}(\Omega)$ is of the form $f_{|\Omega_{\mathbf{i}}} = f_1 + f_2\mathbf{j}$, where the unit vector $\mathbf{j}$ is orthogonal to $\mathbf{i}$, then

$$\begin{aligned}
\sup\{|f(q)| \mid q \in K\} &= \sup\{|f_{|\Omega_{\mathbf{i}}}(q)| \mid q \in K\} \\
&\leq \sup\{|f_1(q)| \mid q \in K\} + \sup\{|f_2(q)| \mid q \in K\} \\
&\leq \alpha_K \left[\int_{\Omega_{\mathbf{i}}} |f_1|^2 \, d\sigma_{\mathbf{i}}\right]^{\frac{1}{2}} + \alpha_K \left[\int_{\Omega_{\mathbf{i}}} |f_2|^2 \, d\sigma_{\mathbf{i}}\right]^{\frac{1}{2}} \\
&\leq 2\alpha_K \left[\int_{\Omega_{\mathbf{i}}} |f_{|\Omega_{\mathbf{i}}}|^2 \, d\sigma_{\mathbf{i}}\right]^{\frac{1}{2}}.
\end{aligned}$$

□

Proposition 2 (Evaluation functional for Ω) *Let $\Omega \subset \mathbb{H}$ be a bounded axially symmetric s-domain. For any compact set $C \subset \Omega$ there exists a constant $\lambda_C > 0$ such that*

$$\sup\{|f(q)| \mid q \in C\} \leq \lambda_C \|f\|_{\mathscr{A}(\Omega)}, \quad \forall f \in \mathscr{A}(\Omega).$$

Proof Let $\mathbf{j} \in \mathbb{S}^2$ and let $K_{\mathbf{j}} := C \cap \mathbb{C}(\mathbf{j})$. Note that $K_{\mathbf{j}}$ is a compact subset of $\Omega_{\mathbf{j}}$, and Proposition 1 gives us a constant $\lambda_{K_{\mathbf{j}}} > 0$ such that

$$\sup\{|f(q)| \mid q \in K_{\mathbf{j}}\} \leq \lambda_{K_{\mathbf{j}}} \|f\|_{\mathscr{A}(\Omega)}, \quad \forall f \in \mathscr{A}(\Omega). \tag{4}$$

Applying the inequality

$$\left|f(x+y\mathbf{i})\right| \leq \left|f(x+y\mathbf{j})\right| + \left|f(x-y\mathbf{j})\right|, \quad \forall x+y\mathbf{i} \in C,\ y \geq 0,$$

which is a direct consequence of (2), one obtains that

$$\sup\{\left|f(q)\right| \mid q \in C\} \leq 2\lambda_{K_\mathbf{j}} \|f\|_{\mathscr{A}(\Omega)}, \quad \forall f \in \mathscr{A}(\Omega).$$

Setting $\lambda_C = 2\lambda_{K_\mathbf{j}}$ the statement follows. □

Theorem 2 *Let $\Omega \subset \mathbb{H}$ be a bounded axially symmetric s-domain. The space $(\mathscr{A}(\Omega), \|\cdot\|_{\mathscr{A}(\Omega)})$ is complete.*

Proof Let $\{f_n\}_{n\in\mathbb{N}} \subset \mathscr{A}(\Omega)$ be a Cauchy sequence, then there exists $\dot{f} \in \mathscr{L}_2(\Omega, \mathbb{H})$ such that

$$\lim_{n\to\infty} \|\dot{f} - f_n\|_{\mathscr{L}_2(\Omega,\mathbb{H})} = 0.$$

Moreover, Proposition 1 implies the following:

(i) There exists the function $\hat{f} : \Omega \to \mathbb{H}$ given by $\hat{f}(q) := \lim_{n\to\infty} f_n(q)$, for any $q \in \Omega$.
(ii) The sequence $\{f_n\}$ converges uniformly to $\hat{f}$ on compact sets. Therefore $\hat{f}$ is a slice regular function on Ω.

Now, for any compact set $K \subset \Omega$ there holds that

$$\begin{aligned} 0 &\leq \int_K |\hat{f} - \dot{f}|^2 d\mu \leq \int_K |\hat{f} - f_n|^2 d\mu + \int_K |\dot{f} - f_n|^2 d\mu \\ &\leq \int_K |\hat{f} - f_n|^2 d\mu + \|\dot{f} - f_n\|^2_{\mathscr{L}_2(\Omega,\mathbb{H})}. \end{aligned}$$

As the right-hand side of the previous inequality tends to zero when $n \to \infty$, one concludes that $\hat{f} = \dot{f} \in \mathscr{A}(\Omega)$. □

4 The Slice Regular Bergman Kernel of the First Kind

We begin this section by noting that Theorem 2 implies that $\mathscr{A}(\Omega)$ is a quaternionic right-linear Hilbert space.

Given any $q \in \Omega$, the evaluation functional $\phi_q : \mathscr{A}(\Omega) \to \mathbb{H}$ is defined by

$$\phi_q[f] := f(q), \quad \forall f \in \mathscr{A}(\Omega).$$

One can prove, using directly Proposition 2, that the evaluation functional ϕ_q is a bounded quaternionic right-linear functional on $\mathscr{A}(\Omega)$.

Even more, Riesz representation theorem for quaternionic right-linear Hilbert space indicates the existence of the unique function $B_q \in \mathscr{A}(\Omega)$ such that

$$\phi_q[f] = \langle B_q, f\rangle_{\mathscr{A}(\Omega)}, \quad \forall f \in \mathscr{A}(\Omega).$$

Denoting $\mathscr{B}(q,\cdot)=\bar{B}_q$, then

$$f(q)=\int_{\Omega}\mathscr{B}(q,\cdot)f\,d\mu,\quad \forall f\in\mathscr{A}(\Omega). \tag{5}$$

Definition 5 The function $\mathscr{B}(\cdot,\cdot):\Omega\times\Omega\to\mathbb{H}$ will be called the Bergman kernel of the first kind associated with Ω.

We recall the following definition that will be useful in the sequel:

Definition 6 A countable family $\{\phi_n\}_{n\in\mathbb{N}}$ in a quaternionic right linear Hilbert space E is called orthonormal total family if the following conditions hold:

- For any $f\in E$ there exists a sequence of quaternions $\{q_n\}_{n\in\mathbb{N}}$ such that
$$f=\sum_{n\geq 0}\phi_n q_n.$$
- $\langle\phi_n,\phi_m\rangle_E=\delta_{n,m}$, for any $n,m\in\mathbb{N}$.
- If $\langle f,\phi_n\rangle_E=0$ for all $n\in\mathbb{N}$, then $f=0$.

Proposition 3 (Properties of the slice regular Bergman kernel of the first kind)

(i) *The slice regular Bergman kernel of the first kind is hermitian;*
$$\mathscr{B}(q,r)=\overline{\mathscr{B}(r,q)},\quad r,q\in\Omega.$$
(ii) *The function $\mathscr{B}(\cdot,\cdot)$ is slice regular in its first coordinate and it is slice right anti-regular in its second coordinate.*

(iii) *The slice regular Bergman kernel of the first kind is the unique function which satisfies the properties* (i), (ii), *and* (5).

(iv) *Let $\{\phi_n\}_{n\in\mathbb{N}}$ be an orthonormal total family of functions in $\mathscr{A}(\Omega)$, and let K be a compact subset of Ω. Then the series*
$$\sum_{n\in\mathbb{N}}\phi(q)\overline{\phi(r)},$$
converges uniformly on $K\times K$ to the Bergman kernel $\mathscr{B}(q,r)$.

Proof

(i)
$$\mathscr{B}(q,r)=\overline{B_q(r)}=\overline{\int_{\Omega}\bar{B}_r B_q\,d\mu}=\int_{\Omega}\bar{B}_q B_r\,d\mu=B_r(q)=\overline{\mathscr{B}(r,q)}.$$

(ii) Since $B_q\in\mathscr{A}(\Omega)$, then
$$\left(\frac{\partial}{\partial x}+\mathbf{i}\frac{\partial}{\partial y}\right)B_{q\,|\Omega_{\mathbf{i}}}(x+y\mathbf{i})=0,\quad \text{on } \Omega_{\mathbf{i}},$$
or equivalently
$$\left(\frac{\partial}{\partial x}+\mathbf{i}\frac{\partial}{\partial y}\right)\mathscr{B}_{|\Omega_{\mathbf{i}}}(x+y\mathbf{i},q)=0,$$

which means that $\mathscr{B}$ is slice regular in its first variable. Finally, applying the quaternionic conjugation we obtain that

$$\mathscr{B}|_{\Omega_{\mathbf{i}}}(q, x + y\mathbf{i})\left(\frac{\partial}{\partial x} - \mathbf{i}\frac{\partial}{\partial y}\right) = 0,$$

which means that $\mathscr{B}$ is slice right anti-regular in its second variable.

(iii) Suppose that the function H satisfies the same properties of $\mathscr{B}$, then

$$\begin{aligned} H(r,q) &= \overline{H(q,r)} = \overline{\int_\Omega \mathscr{B}(q,\cdot)H(\cdot,r)\,d\mu} = \int_\Omega H(r,\cdot)\mathscr{B}(\cdot,q)\,d\mu \\ &= \mathscr{B}(r,q). \end{aligned}$$

(iv) As $\mathscr{B}(\cdot, r) \in \mathscr{A}(\Omega)$, there exists a sequence of quaternions $\{v_n(r)\}_{n\in\mathbb{N}}$, such that

$$\mathscr{B}_\Omega(q,r) = \sum_{n\geq 0} \phi_n(q) v_n(r).$$

Note that for any natural number m we have:

$$\begin{aligned} \overline{\phi_m(r)} &= \overline{\int_\Omega \mathscr{B}(r,q)\phi_m(q)\,d\mu(q)} = \int_\Omega \overline{\phi_m}(q)\mathscr{B}(q,r)\,d\mu(q) \\ &= \int_\Omega \overline{\phi_m}(q)\sum_{n\geq 0}\phi_n(q)v_n(r)\,d\mu(q). \end{aligned}$$

From the uniform convergence on the compact set and from the fact that $\phi_n(q)$ are orthonormal, we get:

$$\overline{\phi_m(r)} = \sum_{n\geq 0}\int_\Omega \overline{\phi_m}(q)\phi_n(q)\,d\mu(q)v_n(r) = v_n(r).$$

Therefore $\mathscr{B}(q,r) = \sum_{n\geq 0}\phi_n(q)\overline{\phi_n(r)}$. □

It is interesting to note that, as in the classical complex case, the kernel $\mathscr{B}(q,r)$ is associated with the solution of a variational problem, as shown in the following proposition:

Proposition 4 *Let $u \in \Omega$ be fixed and define*

$$F_{u,1} := \{ f \in \mathscr{A}(\Omega) \mid f(u) = 1 \}.$$

Then the function

$$f^*(q) := \frac{\mathscr{B}(q,u)}{\mathscr{B}(u,u)}, \quad \forall q \in \Omega,$$

belongs to $F_{u,1}$, and it is the unique solution of the variational problem of finding

$$\inf\{ \|f\|_{\mathscr{A}(\Omega)} \mid f \in F_{u,1} \} = \|f^*\|_{\mathscr{A}(\Omega)}.$$

Proof The proof is clear from the following facts:

(i) Hölder's inequality and $\|\mathscr{B}(\cdot,u)\|_{\mathscr{A}(\Omega)}=0$ implies that

$$f(u)=0, \quad \forall f\in\mathscr{A}(\Omega).$$

Therefore $\|\mathscr{B}(\cdot,u)\|_{\mathscr{A}(\Omega)}>0$, and

$$\mathscr{B}(u,u)=\int_\Omega \mathscr{B}(u,\xi)\mathscr{B}(\xi,u)\,d\xi=\left\|\mathscr{B}(\cdot,u)\right\|^2_{\mathscr{A}(\Omega)}>0.$$

(ii)
$$\left\|\frac{\mathscr{B}(\cdot,u)}{\mathscr{B}(u,u)}\right\|_{\mathscr{A}(\Omega)}=\frac{1}{\|\mathscr{B}(\cdot,u)\|_{\mathscr{A}(\Omega)}}.$$

(iii) Take $f\in F_{u,1}$, then

$$1=f(u)=\int_\Omega B(u,\xi)f(\xi)\,d\xi\le\left\|\mathscr{B}(\cdot,u)\right\|_{\mathscr{A}(\Omega)}\|f\|_{\mathscr{A}(\Omega)}.$$ □

As a direct consequence of Proposition 3, one can find an explicit representation of the slice regular Bergman kernel of the first kind in the unit ball.

Corollary 1 *The slice regular Bergman kernel of the first kind for the unit ball* $\mathbb{B}^4$ *is given by*:

$$\mathscr{B}(q,r)=\sum_{n\ge0}\frac{\Phi_n(q,r)}{\mathscr{M}_n(q,r)}+\sum_{n\ge0}\frac{\Psi_n(q,r)}{\mathscr{N}_n(q,r)},$$

where

$$\Phi_n(q,r):=q^{2n+1}(\bar r)^{2n+1}-8\pi^4\sum_{k=0}^{n}\left(q^{2k+1}(\bar r)^{2n+1}+q^{2n+1}(\bar r)^{2k+1}\right)\alpha_{n,k}$$
$$+64\pi^8\sum_{k,\ell=0}^{n}q^{2k+1}(\bar r)^{2\ell+1}\alpha_{n,k}\alpha_{n,\ell},$$

$$\mathscr{M}_n:=\frac{\pi^2}{2n+3}-16\pi^6\sum_{\ell=0}^{n}\frac{(2(n-\ell))!\alpha_{n,\ell}\lambda_{n,\ell}}{(n+\ell+3)(n-\ell+1)!2^{2(n-\ell)}}$$
$$+128\pi^{10}\sum_{\substack{k>\ell\\k,\ell=0}}^{n}\frac{(2(k-\ell))!\alpha_{n,k}\alpha_{n,\ell}\lambda_{k,\ell}}{(k+\ell+3)(k-\ell+1)!2^{2(k-\ell)}}$$
$$+64\pi^{10}\sum_{k=0}^{n}\frac{\alpha_{n,k}^2}{2k+3},$$

$$\Psi_n(q,r):=q^{2n}(\bar r)^{2n}-8\pi^4\sum_{k=0}^{n}\left(q^{2k}(\bar r)^{2n}+q^{2n}(\bar r)^{2k}\right)\beta_{n,k}$$
$$+64\pi^8\sum_{k,\ell=0}^{n}q^{2k}(\bar r)^{2\ell}\beta_{n,k}\beta_{n,\ell},$$

$$\mathcal{N}_n(q,r) := \frac{\pi^2}{2n+2} - 16\pi^6 \sum_{\ell=0}^{n} \frac{(2(n-\ell))!\beta_{n,\ell}\lambda_{n,\ell}}{(n+\ell+2)(n-\ell+1)!2^{2(n-\ell)}}$$
$$+128\pi^{10} \sum_{\substack{k>\ell\\k,\ell=0}}^{n} \frac{(2(k-\ell))!\beta_{n,k}\beta_{n,\ell}\lambda_{k,\ell}}{(k+\ell+2)(k-\ell+1)!2^{2(k-\ell)}} + 32\pi^{10}\sum_{k=0}^{n}\frac{\beta_{n,k}^2}{k+1},$$
$$\alpha_{n,k} := \frac{(2(n-k))!\lambda_{n,k}}{2(n+k+3)(4k+6)(n-k+1)!2^{2(n-k)+1}},$$
$$\beta_{n,k} := \frac{(2(n-k))!\lambda_{n,k}}{2(n+k+2)(4k+4)(n-k+1)!2^{2(n-k)+1}},$$

and

$$\lambda_{n,k} := \sum_{m=0}^{n-k} \frac{(-1)^m(2m+1)}{m!(n-k-m)!}.$$

Proof One can see directly that the family of functions $\{q^n\}_{n\in\mathbb{N}}$ is not an orthonormal family. Nevertheless, the paper [9] tells us that this family generates the set of slice regular functions. Thus, the family $\{q^n\}_{n\in\mathbb{N}}$ is contained in $\mathcal{A}(\mathbb{B}^4)$ and generates any element of $\mathcal{A}(\mathbb{B}^4)$. The orthonormalization process gives us the following:

$$u_1(q) = q,$$
$$u_2(q) = q^2 - q\langle q,q\rangle^{-1}_{\mathcal{A}(\mathbb{B}^4)}\langle q,q^2\rangle_{\mathcal{A}(\mathbb{B}^4)},$$
$$u_3(q) = q^3 - q^2\langle q^2,q^2\rangle^{-1}_{\mathcal{A}(\mathbb{B}^4)}\langle q^2,q^3\rangle_{\mathcal{A}(\mathbb{B}^4)} - q\langle q,q\rangle^{-1}_{\mathcal{A}(\mathbb{B}^4)}\langle q,q^3\rangle_{\mathcal{A}(\mathbb{B}^4)},$$
$$\vdots$$

Therefore, $\{\frac{u_i}{\|u_i\|_{\mathcal{A}(\mathbb{B}^4)}}\}_{i\in\mathbb{N}}$ is an orthonormal total family contained in $\mathcal{A}(\mathbb{B}^4)$.

Moreover, let $n,m\in\mathbb{N}$, the following equalities are obtained passing to the spherical coordinates:

(i)
$$\langle q^{2n},q^{2m+1}\rangle_{\mathcal{A}(\mathbb{B}^4)} = 0, \quad \forall n,m\in\mathbb{N}.$$

(ii)
$$\langle q^n,q^n\rangle_{\mathcal{A}(\mathbb{B}^4)} = \frac{2\pi^2}{2n+4}, \quad \forall n\in\mathbb{N}.$$

(iii)
$$\langle q^{2n},q^{2m}\rangle_{\mathcal{A}(\mathbb{B}^4)} = \frac{4\pi^4(2(n-m))!\lambda_{n,m}}{2(n+m+2)2^{2(n-m)+1}(n-m+1)!},$$
$$\forall n,m\in\mathbb{N}, \text{ with } n>m.$$

(iv)
$$\langle q^{2n+1},q^{2m+1}\rangle_{\mathcal{A}(\mathbb{B}^4)} = \frac{4\pi^4(2(n-m))!\lambda_{n,m}}{2(n+m+3)2^{2(n-m)+1}(n-m+1)!},$$
$$\forall n,m\in\mathbb{N}, \text{ with } n>m.$$

The above identities allow one to get that

$$u_{2n+1}(q)=q^{2n+1}-8\pi^4\sum_{k=0}^{n}q^{2k+1}\frac{(2(n-k))!\lambda_{n,k}}{(2n+2k+6)(4k+6)(n-k+1)!2^{2(n-k)+1}},$$

$$u_{2n}(q)=q^{2n}-8\pi^4\sum_{k=0}^{n}q^{2k}\frac{(2(n-k))!\lambda_{n,k}}{(2n+2k+4)(4k+4)(n-k+1)!2^{2(n-k)+1}},$$

$$\begin{aligned}\|u_{2n+1}\|^2_{\mathscr{A}(\mathbb{B}^4)}&=\frac{\pi^2}{2n+3}-16\pi^6\sum_{\ell=0}^{n}\frac{(2(n-\ell))!\alpha_{n,\ell}\lambda_{n,\ell}}{(n+\ell+3)(n-\ell+1)!2^{2(n-\ell)}}\\&+128\pi^{10}\sum_{\substack{k>\ell\\k,\ell=0}}^{n}\frac{(2(k-\ell))!\alpha_{n,k}\alpha_{n,\ell}\lambda_{k,\ell}}{(k+\ell+3)(k-\ell+1)!2^{2(k-\ell)}}\\&+64\pi^{10}\sum_{k=0}^{n}\frac{\alpha^2_{n,k}}{2k+3},\end{aligned}$$

and

$$\begin{aligned}\|u_{2n}\|^2_{\mathscr{A}(\mathbb{B}^4)}&=\frac{\pi^2}{2n+2}-16\pi^6\sum_{\ell=0}^{n}\frac{(2(n-\ell))!\beta_{n,\ell}\lambda_{n,\ell}}{(n+\ell+2)(n-\ell+1)!2^{2(n-\ell)}}\\&+128\pi^{10}\sum_{\substack{k>\ell\\k,\ell=0}}^{n}\frac{(2(k-\ell))!\beta_{n,k}\beta_{n,\ell}\lambda_{k,\ell}}{(k+\ell+2)(k-\ell+1)!2^{2(k-\ell)}}\\&+32\pi^{10}\sum_{k=0}^{n}\frac{\beta^2_{n,k}}{k+1}.\end{aligned}$$

□

5 The Slice Regular Bergman Theory of the Second Kind

Let $\Omega\subseteq\mathbb{H}$ be an axially symmetric s-domain. We introduce here a family of Bergman-type spaces. For every $\mathbf{i}\in\mathbb{S}^2$ we set

$$\mathscr{A}(\Omega_{\mathbf{i}}):=\left\{f\in\mathscr{SR}(\Omega)\;\middle|\;\|f\|^2_{\mathscr{A}(\Omega_{\mathbf{i}})}:=\int_{\Omega_{\mathbf{i}}}|f_{|\Omega_{\mathbf{i}}}|^2\,d\sigma_{\mathbf{i}}<\infty\right\}.$$

On $\mathscr{A}(\Omega_{\mathbf{i}})$ we define the scalar product

$$\langle f,g\rangle_{\mathscr{A}(\Omega_{\mathbf{i}})}=\int_{\Omega_{\mathbf{i}}}\overline{f}\,d\sigma_{\mathbf{i}}\,g.$$

It is useful to compare this with formula (3) and Definition 4, where the functions are also slice regular but, besides, they are a priori in $\mathscr{L}_2(\Omega,\mathbb{H})$ while here only the restrictions of slice regular functions are in the corresponding $\mathscr{L}_2(\Omega_{\mathbf{i}},\mathbb{H})$.

In this section there are a few results that can be proved as in the complex case. To point out this fact, we simply write that the proofs follow by standard arguments.

Theorem 3 (Completeness of $\mathscr{A}(\Omega_{\mathbf{i}})$) *Let Ω be a bounded axially symmetric s-domain. The spaces $(\mathscr{A}(\Omega_{\mathbf{i}}), \|\cdot\|_{\mathscr{A}(\Omega_{\mathbf{i}})})$ are complete for every $\mathbf{i} \in \mathbb{S}^2$.*

Proof It follows by standard arguments as in the complex case. □

Remark 4 Let Ω be a given bounded axially symmetric s-domain in $\mathbb{H}$.

- Theorem 3 implies that the Bergman spaces $\mathscr{A}(\Omega_{\mathbf{i}})$, for $\mathbf{i} \in \mathbb{S}^2$, are quaternionic right linear Hilbert spaces.
- Since $f_{|\Omega_{\mathbf{i}}} = F + G\mathbf{j}$ where $F, G : \Omega_{\mathbf{i}} \to \mathbb{C}(\mathbf{i})$ are holomorphic functions and $\mathbf{j} \perp \mathbf{i}$, then for $f_{|\Omega_{\mathbf{i}}} \in \mathscr{A}(\Omega_{\mathbf{i}})$ there holds: $F, G \in \mathscr{A}(\Omega_{\mathbf{i}})$. This follows from the inequalities

$$\left.\begin{array}{l} \int_{\Omega_{\mathbf{i}}} |F|^2 \, d\sigma_{\mathbf{i}} \\ \int_{\Omega_{\mathbf{i}}} |G|^2 \, d\sigma_{\mathbf{i}} \end{array}\right\} \leq \|f_{|\Omega_{\mathbf{i}}}\|^2_{\mathscr{A}(\Omega_{\mathbf{i}})}.$$

Proposition 1 and Remark 4 imply that given any $q \in \Omega_{\mathbf{i}}$ the evaluation functional $\phi_q : \mathscr{A}(\Omega_{\mathbf{i}}) \to \mathbb{H}$, given by

$$\phi_q[f] := f(q), \quad \forall f \in \mathscr{A}(\Omega_{\mathbf{i}}),$$

is a bounded quaternionic right-linear functional on $\mathscr{A}(\Omega_{\mathbf{i}})$ for every $\mathbf{i} \in \mathbb{S}^2$.

The Riesz representation theorem for quaternionic right-linear Hilbert spaces shows the existence of the unique function $K_q(\mathbf{i}) \in \mathscr{A}(\Omega_{\mathbf{i}})$ such that

$$\phi_q[f_{|\Omega_{\mathbf{i}}}] = \langle K_q(\mathbf{i}), f_{|\Omega_{\mathbf{i}}} \rangle_{\mathscr{A}(\Omega_{\mathbf{i}})}.$$

Denoting $\mathscr{K}_{\Omega_{\mathbf{i}}}(q, \cdot) := \overline{K_q}(\mathbf{i})$ we have

$$f_{|\Omega_{\mathbf{i}}}(q) = \int_{\Omega_{\mathbf{i}}} \mathscr{K}_{\Omega_{\mathbf{i}}}(q, \cdot) \, d\sigma_{\mathbf{i}} \, f_{|\Omega_{\mathbf{i}}}, \quad \forall f_{|\Omega_{\mathbf{i}}} \in \mathscr{A}(\Omega_{\mathbf{i}}), \tag{6}$$

moreover, we give the following

Definition 7 (Bergman kernel associated with $\Omega_{\mathbf{i}}$) The function

$$\mathscr{K}_{\Omega_{\mathbf{i}}}(\cdot,\cdot) : \Omega_{\mathbf{i}} \times \Omega_{\mathbf{i}} \to \mathbb{H}$$

will be called the Bergman kernel associated with $\Omega_{\mathbf{i}}$.

The Bergman kernel $\mathscr{K}_{\Omega_{\mathbf{i}}}$ is, in general, a quaternionic valued function which however satisfies the Cauchy-Riemann equation if restricted to any complex plane $\mathbb{C}(\mathbf{i})$. Thus one may ask if this function is related to the classical Bergman kernel. The answer is contained in the next result.

Proposition 5 *The Bergman kernel $\mathscr{K}_{\Omega_{\mathbf{i}}}$ coincides with the classical Bergman kernel on $\Omega_{\mathbf{i}}$, for any $\mathbf{i} \in \mathbb{S}$.*

Proof By the Splitting Lemma, $f_{|\Omega_{\mathbf{i}}} = F + G\mathbf{j}$, where F, G are $\mathbb{C}(\mathbf{i})$-valued holomorphic functions, so we have

$$\begin{aligned}\phi_q[f_{|\Omega_{\mathbf{i}}}] &= \langle K_q(\mathbf{i}), F + G\mathbf{j}\rangle_{\mathscr{A}(\Omega_{\mathbf{i}})}\\ &= \langle K_q(\mathbf{i}), F\rangle_{\mathscr{A}(\Omega_{\mathbf{i}})} + \langle K_q(\mathbf{i}), G\rangle_{\mathscr{A}(\Omega_{\mathbf{i}})}\mathbf{j}, \quad \forall f_{|\Omega_{\mathbf{i}}} \in \mathscr{A}(\Omega_{\mathbf{i}}).\end{aligned}$$

If $q \in \mathbb{C}(\mathbf{i})$ we have

$$\begin{aligned}F(q) &= \int_{\Omega_{\mathbf{i}}} \mathscr{K}_{\Omega_{\mathbf{i}}}(q, \cdot)\, d\sigma_{\mathbf{i}}\, F,\\ G(q) &= \int_{\Omega_{\mathbf{i}}} \mathscr{K}_{\Omega_{\mathbf{i}}}(q, \cdot)\, d\sigma_{\mathbf{i}}\, G.\end{aligned}$$

However, on the complex plane $\mathbb{C}(\mathbf{i})$ the holomorphic functions F, G can be written using the classical Bergman kernel in one complex variable. By the uniqueness of the function $\mathscr{K}_{\Omega_{\mathbf{i}}}(q, \cdot)$, it coincides with the classical Bergman kernel. □

As an immediate consequence of the previous proposition, we have the following

Remark 5 (Properties of the Bergman kernel associated with $\Omega_{\mathbf{i}}$) Let $\Omega \subset \mathbb{H}$ be an axially symmetric s-domain. Let $\mathbf{i} \in \mathbb{S}^2$ and $\Omega_{\mathbf{i}} := \Omega \cap \mathbb{C}(\mathbf{i})$. Then the slice Bergman kernel associated to $\Omega_{\mathbf{i}}$ satisfies the properties:

(i) $\mathscr{K}_{\Omega_{\mathbf{i}}}(z, w) = \overline{\mathscr{K}_{\Omega_{\mathbf{i}}}(w, z)}$, for $z, w \in \Omega_{\mathbf{i}}$.
(ii) $\mathscr{K}_{\Omega_{\mathbf{i}}}(\cdot,\cdot)$ is anti-holomorphic with respect to its second coordinate.
(iii) $\mathscr{K}_{\Omega_{\mathbf{i}}}(\cdot,\cdot)$ is a reproducing kernel on $\Omega_{\mathbf{i}}$.
(iv) Let $\{\phi_n\}_{n\in\mathbb{N}}$ be a orthonormal total family of functions in $A(\Omega_{\mathbf{i}})$, and let $K_{\mathbf{i}}$ be a compact subset of $\Omega_{\mathbf{i}}$. Then

$$\sum_{n\geq 0} \phi(z)\overline{\phi(w)}$$

sums uniformly on $K_{\mathbf{i}} \times K_{\mathbf{i}}$ to the Bergman kernel $\mathscr{K}_{\Omega_{\mathbf{i}}}(z, w)$.

The proofs of the various facts are immediate consequences of Proposition 5, but a comment is in order. The function $\mathscr{K}_{\Omega_{\mathbf{i}}}(z, w)$ is $\mathbb{C}(\mathbf{i})$-valued. The conjugation considered in (i) is the quaternionic conjugation which, in particular, restricts to the complex conjugation on each complex plane, so (i) follows.

Proposition 6 *Let Ω be an axially symmetric s-domain and consider two imaginary units $\mathbf{t}, \mathbf{s} \in \mathbb{S}$. Let us consider the function defined by*

$$\mathscr{K}_{\Omega}(x + y\mathbf{i}, r) := \frac{1}{2}(1 - \mathbf{it})\mathscr{K}_{\Omega_{\mathbf{t}}}(x + y\mathbf{t}, r) + \frac{1}{2}(1 + \mathbf{it})\mathscr{K}_{\Omega_{\mathbf{t}}}(x - y\mathbf{t}, r). \quad (7)$$

Then it can also be written as

$$\mathscr{K}_{\Omega}(x + y\mathbf{i}, r) = \frac{1}{2}(1 - \mathbf{is})\mathscr{K}_{\Omega_{\mathbf{s}}}(x + y\mathbf{s}, r) + \frac{1}{2}(1 + \mathbf{is})\mathscr{K}_{\Omega_{\mathbf{s}}}(x - y\mathbf{s}, r). \quad (8)$$

Proof Both functions defined by the right-hand sides of (7) and (8) are slice regular, by construction, and coincide with the Bergman kernel $\mathscr{K}_{\Omega_{\mathbf{i}}}(z,w)$ on the complex plane $\mathbb{C}(\mathbf{i})$. Thus, by the identity principle they coincide on Ω. □

This result assures that the following definition is well posed.

Definition 8 (Slice Bergman kernel of the second kind associated with Ω) Let $\mathscr{K}_{\Omega_{\mathbf{j}}}(\cdot,\cdot):\Omega_{\mathbf{j}}\times\Omega_{\mathbf{j}}\to\mathbb{H}$ be the Bergman kernel associated with $\Omega_{\mathbf{j}}$. We will call slice Bergman kernel of the second kind associated with Ω the function

$$\mathscr{K}_{\Omega}:\Omega\times\Omega\to\mathbb{H},$$

defined by

$$\mathscr{K}_{\Omega}(x+y\mathbf{i},r):=\frac{1}{2}(1-\mathbf{ij})\mathscr{K}_{\Omega_{\mathbf{j}}}(x+y\mathbf{j},r)+\frac{1}{2}(1+\mathbf{ij})\mathscr{K}_{\Omega_{\mathbf{j}}}(x-y\mathbf{j},r).$$

Proposition 7 (Properties of the slice Bergman kernel of the second kind associated with Ω) *Let $\Omega\subset\mathbb{H}$ be an axially symmetric s-domain. Then the slice Bergman kernel of the second kind $\mathscr{K}_{\Omega}$ satisfies the properties*:

(i) $\mathscr{K}_{\Omega}(q,r)=\overline{\mathscr{K}_{\Omega}(r,q)}$, *for $r,q\in\Omega$.*
(ii) $\mathscr{K}_{\Omega}(\cdot,\cdot)$ *is slice right anti-regular with respect to its second variable.*
(iii) $\mathscr{K}_{\Omega}(\cdot,\cdot)$ *is a reproducing kernel on Ω. More precisely, let $\mathbf{i}\in\mathbb{S}^2$ then*

$$f(q)=\int_{\Omega\cap\mathbb{C}(\mathbf{i})}\mathscr{K}_{\Omega}(q,\cdot)f\,d\sigma_{\mathbf{i}},$$

and the integral does not depend on $\mathbf{i}\in\mathbb{S}^2$.

Proof We have the following identities:

$$\begin{aligned}\overline{\mathscr{K}_{\Omega}(x+y\mathbf{i},r)}&=\overline{\frac{1}{2}(1-\mathbf{ij})\mathscr{K}_{\Omega_{\mathbf{j}}}(x+y\mathbf{j},r)+\frac{1}{2}(1+\mathbf{ij})\mathscr{K}_{\Omega_{\mathbf{j}}}(x-y\mathbf{j},r)}\\&=\overline{\frac{1}{2}(1-\mathbf{ij})\mathscr{K}_{\Omega_{\mathbf{j}}}(x+y\mathbf{j},r)}+\overline{\frac{1}{2}(1+\mathbf{ij})\mathscr{K}_{\Omega_{\mathbf{j}}}(x-y\mathbf{j},r)}\\&=\overline{\mathscr{K}_{\Omega_{\mathbf{j}}}(x+y\mathbf{j},r)}\,\overline{\frac{1}{2}(1-\mathbf{ij})}+\overline{\mathscr{K}_{\Omega_{\mathbf{j}}}(x-y\mathbf{j},r)}\,\overline{\frac{1}{2}(1+\mathbf{ij})}\\&=\mathscr{K}_{\Omega_{\mathbf{j}}}(r,x+y\mathbf{j})\frac{1}{2}(1+\mathbf{ij})+\mathscr{K}_{\Omega_{\mathbf{j}}}(r,x-y\mathbf{j})\frac{1}{2}(1-\mathbf{ij}).\end{aligned}\tag{9}$$

Since (9) is the right hand side of the representation formula for slice right anti-regular functions, we get:

$$\overline{\mathscr{K}_{\Omega}(x+y\mathbf{i},r)}=\mathscr{K}_{\Omega}(r,x+y\mathbf{i}),$$

which is equivalent to the formula (i).

(ii) The fact that the kernel $\mathscr{K}_{\Omega}(q,r)$ is slice left regular with respect to q and slice right anti-regular with respect to r follows from the representation formula and Remark 5, (ii).

(iii) The integral

$$f(q) = \int_{\Omega \cap \mathbb{C}(\mathbf{i})} \mathscr{K}_\Omega(q, \cdot) f \, d\sigma_{\mathbf{i}}$$

does not depend on $\mathbf{i} \in \mathbb{S}^2$ by the representation formula. □

We conclude by noting that the slice Bergman kernel of the second kind $\mathscr{K}_\Omega$ can be computed in closed form whenever it is known the corresponding slice Bergman kernel $\mathscr{K}_{\Omega_{\mathbf{i}}}$. For example, it is possible to construct $\mathscr{K}_\Omega$ when Ω is the unit ball in $\mathbb{H}$ thus overcoming the difficulties shown in Corollary 1. The knowledge of a closed form of the kernel is useful for several reasons and has been used in [12] to construct the Bergman-Sce transform in the setting of slice monogenic functions. We will postpone to a forthcoming paper a more detailed study of the slice Bergman kernel of the second kind as well as the comparison of the two theories of first and second kind.

References

1. Bergman, S.: The Kernel Function and Conformal Mapping. Am. Math. Soc., Providence (1970)
2. Bergman, S., Schiffer, M.: Kernel Functions and Elliptic Differential Equations in Mathematical Physics. Academic Press, New York (1953)
3. Brackx, F., Delanghe, R.: Hypercomplex function theory and Hilbert modules with reproducing kernel. Proc. Am. Math. Soc. **37**, 545–576 (1978)
4. Brackx, F., Delanghe, R., Sommen, F.: Clifford Analysis. Pitman Research Notes in Mathematics, vol. 76 (1982)
5. Colombo, F., Sabadini, I.: A structure formula for slice monogenic functions and some of its consequences. In: Hypercomplex Analysis. Trends in Mathematics, pp. 101–114. Birkhäuser, Basel (2009)
6. Colombo, F., Sabadini, I.: The Cauchy formula with s-monogenic kernel and a functional calculus for noncommuting operators. J. Math. Anal. Appl. **373**, 655–679 (2011)
7. Colombo, F., Gentili, G., Sabadini, I., Struppa, D.C.: Extension results for slice regular functions of a quaternionic variable. Adv. Math. **222**, 1793–1808 (2009)
8. Colombo, F., Sabadini, I., Struppa, D.C.: Slice monogenic functions. Isr. J. Math. **171**, 385–403 (2009)
9. Colombo, F., Gentili, G., Sabadini, I.: A Cauchy kernel for slice regular functions. Ann. Glob. Anal. Geom. **37**, 361–378 (2010)
10. Colombo, F., Sabadini, I., Sommen, F.: The Fueter mapping theorem in integral form and the $\mathscr{F}$-functional calculus. Math. Methods Appl. Sci. **33**, 2050–2066 (2010)
11. Colombo, F., Sabadini, I., Struppa, D.C.: An extension theorem for slice monogenic functions and some of its consequences. Isr. J. Math. **177**, 369–389 (2010)
12. Colombo, F., González-Cervantes, J.O., Sabadini, I.: The Bergman-Sce transform for slice monogenic functions. Math. Methods Appl. Sci. **34**, 1896–1909 (2011)
13. Constales, D.: The Bergman and Szegö kernels for separately monogenic functions. Z. Anal. Anwend. **9**, 97–103 (1990)
14. Constales, D., Kraußhar, R.S.: Bergman kernels for rectangular domains and multiperiodic functions in Clifford analysis. Math. Methods Appl. Sci. **25**, 1509–1526 (2002)
15. Constales, D., Kraußhar, R.S.: Bergman spaces of higher-dimensional hyperbolic polyhedron-type domains I. Math. Methods Appl. Sci. **29**, 85–98 (2006)

16. Delanghe, R.: On Hilbert modules with reproducing kernel. In: Functional and Theoretical Methods: Partial Differential Equations, Proceedings of the International Symposium, Darmstadt, 1976. Lecture Notes in Mathematics, vol. 561, pp. 158–170 (1976)
17. Gentili, G., Struppa, D.C.: A new theory of regular functions of a quaternionic variable. Adv. Math. **216**, 279–301 (2007)
18. González-Cervantes, J.O., Luna-Elizarrarás, M.E., Shapiro, M.: On some categories and functors in the theory of quaternionic Bergman spaces. Adv. Appl. Clifford Algebras **19**(2), 325–338 (2009)
19. Krantz, S.G.: Function Theory of Several Complex Variables, 2nd edn. Wadsworth & Brooks, Belmont (1982)
20. Shapiro, M., Vasilevski, N.: On the Bergman kernel function in hyperholomorphic analysis. Acta Appl. Math. **46**, 1–27 (1977)
21. Shapiro, M., Vasilevski, N.: On the Bergman kernel function in Clifford analysis. In: Bracks, F., et al. (eds.) Clifford Analysis and Their Applications in Mathematical Physics, Proceedings of the Third Conference, Deinze, Belgium, 1993. Fundamental Theories of Physics, vol. 55, pp. 183–192. Kluwer Academic, Dordrecht (1993)
22. Shapiro, M., Vasilevski, N.: On the Bergman kernel functions in quaternionic analysis. Russ. Math. (Izv. VUZ) **42**(2), 81–85 (1998)

A Bloch-Landau Theorem for Slice Regular Functions

Chiara Della Rocchetta, Graziano Gentili, and Giulia Sarfatti

Abstract The Bloch-Landau Theorem is one of the basic results in the geometric theory of holomorphic functions. It establishes that the image of the open unit disc $\mathbb{D}$ under a holomorphic function f (such that $f(0) = 0$ and $f'(0) = 1$) always contains an open disc with radius larger than a universal constant. In this paper we prove a Bloch-Landau type Theorem for slice regular functions over the skew field $\mathbb{H}$ of quaternions. If f is a regular function on the open unit ball $\mathbb{B} \subset \mathbb{H}$, then for every $w \in \mathbb{B}$ we define the *regular translation* $\tilde{f}_w$ of f. The peculiarities of the non commutative setting lead to the following statement: there exists a universal open set contained in the image of $\mathbb{B}$ through some regular translation $\tilde{f}_w$ of any slice regular function $f : \mathbb{B} \to \mathbb{H}$ (such that $f(0) = 0$ and $\partial_C f(0) = 1$). For technical reasons, we introduce a new norm on the space of regular functions on open balls centred at the origin, equivalent to the uniform norm, and we investigate its properties.

1 Introduction

After the great development of the theory of holomorphic functions, there have been many attempts, and successes of different character, to build analogous theories of regular functions whose domain and range were the quaternions $\mathbb{H}$. The most successful theory, by far, has been the one due to Fueter (see the foundation papers [7, 8], the nice survey [21], the book [1] and references therein). More recently, Cullen (see [5]) gave a definition of regularity that Gentili and Struppa reinterpreted and developed in [12, 13], giving rise to the rich theory of slice regular functions. For this class of functions there have been proved the corresponding of several results of the complex setting, that often assume a different flavour in the quaternionic

C. Della Rocchetta · G. Gentili (✉) · G. Sarfatti
Department of Mathematics "U. Dini", University of Florence, Viale Morgagni 67/A, 50134 Florence, Italy
e-mail: gentili@math.unifi.it

C. Della Rocchetta
e-mail: mchidel@gmail.com

G. Sarfatti
e-mail: sarfatti@math.unifi.it

G. Gentili et al. (eds.), *Advances in Hypercomplex Analysis*, Springer INdAM Series 1, DOI 10.1007/978-88-470-2445-8_4, © Springer-Verlag Italia 2013

setting. For instance we can cite, among the most basic, results that concern the Cauchy Representation Formula and the Cauchy kernel, the Identity Principle, the Maximum Modulus Principle, the Open Mapping Theorem and a new notion of analyticity [3, 9–11, 13]. Moreover, the theory of slice regular functions has been extended and generalized to the Clifford Algebras setting, and to the more general setting of alternative algebras, originating a class of functions that is also under deep investigation at the moment (see [4] and references therein, and [15, 16]).

An important property, that distinguishes slice regular functions from the Fueter regular ones, is that power series with quaternionic coefficients on the right, $\sum_{n=0}^{\infty} q^n a_n$ are slice regular. Furthermore, on open balls centred at the origin, to have a power series expansion is a necessary and sufficient condition for a function to be slice regular. Since we will work with functions defined on open balls centred at the origin, we will make use of this characterization. Indeed, the purpose of this paper is to prove an analog of the Bloch-Landau Theorem for slice regular functions. In the complex case, this result is an important fact in the study of the range of holomorphic functions defined on the open unit disc $\mathbb{D}$. It states that the image of $\mathbb{D}$ through a holomorphic function cannot be "too much thin". In fact, under certain normalizations, it contains always a disk with a universal radius. One of the first lower bounds, $\frac{1}{16}$, of this constant is due to Landau, see the book [19]. The same author gave also better estimates in [18]. Recently, the Bloch's Theorem in the quaternionic setting has been investigated with success in [17].

In our approach, since composition of slice regular functions is not regular in general, we have to define the set of *regular translations* of a regular function f defined on the open unit ball $\mathbb{B}$ of the space of quaternions. We then prove the existence of a universal open set $\mathscr{O}$, different from a ball, always contained in the image of the open unit ball $\mathbb{B}$ under some regular translation of any (normalized) slice regular function f. Thus we provide a further tool to the geometric theory of slice regular functions.

The paper is organized as follows: Sect. 2 is dedicated to set up the notation and to give the preliminary results; in Sect. 3 we prove an important property of the uniform norm on balls centred at the origin; in Sect. 4 we define a new norm on the space of slice regular functions (on open balls centred at the origin) and we use it to set up a mean value theorem; the last section is devoted to prove a Bloch-Landau type Theorem for slice regular functions.

2 Preliminaries

Let $\mathbb{H}$ denote the non commutative real algebra of quaternions with standard basis $\{1, i, j, k\}$. The elements of the basis satisfy the multiplication rules

$$i^2 = j^2 = k^2 = -1, \qquad ij = k = -ji, \qquad jk = i = -kj, \qquad ki = j = -ik,$$

that extend by distributivity to all $q = x_0 + x_1 i + x_2 j + x_3 k$ in $\mathbb{H}$. Every element of this form is composed by the *real* part $\mathrm{Re}(q) = x_0$ and the *imaginary* part

$\mathrm{Im}(q) = x_1 i + x_2 j + x_3 k$. The *conjugate* of $q \in \mathbb{H}$ is then $\bar{q} = \mathrm{Re}(q) - \mathrm{Im}(q)$ and its *modulus* is defined as $|q|^2 = q\bar{q} = \mathrm{Re}(q)^2 + |\mathrm{Im}(q)|^2$. We can therefore calculate the multiplicative inverse of each $q \neq 0$ as $q^{-1} = \frac{\bar{q}}{|q|^2}$. Notice that for all non real $q \in \mathbb{H}$, the quantity $\frac{\mathrm{Im}(q)}{|\mathrm{Im}(q)|}$ is an imaginary unit, that is a quaternion whose square equals -1. Then we can express every $q \in \mathbb{H}$ as $q = x + yI$, where x, y are real (if $q \in \mathbb{R}$, then $y = 0$) and I is an element of the unit 2-dimensional sphere of purely imaginary quaternions,

$$\mathbb{S} = \{q \in \mathbb{H} \mid q^2 = -1\}.$$

In the sequel, for every $I \in \mathbb{S}$ we will define L_I to be the plane $\mathbb{R} + \mathbb{R}I$, isomorphic to $\mathbb{C}$, and, if Ω is a subset of $\mathbb{H}$, by Ω_I the intersection $\Omega \cap L_I$. Also, for $R > 0$, we will denote the open ball centred at the origin with radius R by

$$B(0, R) = \{q \in \mathbb{H} \mid |q| < R\}.$$

We can now recall the definition of slice regularity.

Definition 1 A function $f : B = B(0, R) \to \mathbb{H}$ is *slice regular* if, for all $I \in \mathbb{S}$, its restriction f_I to B_I is *holomorphic*, that is it has continuous partial derivatives and it satisfies

$$\overline{\partial_I} f(x + yI) := \frac{1}{2}\left(\frac{\partial}{\partial x} + I\frac{\partial}{\partial y}\right) f_I(x + yI) = 0$$

for all $x + yI \in B_I$.

In the sequel, we will avoid the prefix slice when referring to slice regular functions. For regular functions we can give the following natural definition of derivative.

Definition 2 Let $f : B(0, R) \to \mathbb{H}$ be a regular function. The *slice derivative* (or *Cullen derivative*) of f at $q = x + yI$ is defined as

$$\partial_C f(x + yI) = \frac{\partial}{\partial x} f(x + yI).$$

We remark that this definition is well posed because it is applied only to regular functions. Moreover, since the operators ∂_C and $\overline{\partial_I}$ do commute, the slice derivative of a regular function is still regular. Hence, we can iterate the differentiation obtaining (see for instance [13]),

$$\partial_C^n f = \frac{\partial^n}{\partial x^n} f \quad \text{for any } n \in \mathbb{N}.$$

As stated in [13], a quaternionic power series $\sum_{n \geq 0} q^n a_n$ with $\{a_n\}_{n\in\mathbb{N}} \subset \mathbb{H}$ defines a regular function in its domain of convergence, which is a ball $B(0, R)$ with R equal to the radius of convergence of the power series. Moreover, in [13], it is also proved that

Theorem 1 *A function f is regular on $B = B(0, R)$ if and only if f has a power series expansion*

$$f(q) = \sum_{n \geq 0} q^n a_n \quad \text{with } a_n = \frac{1}{n!}\frac{\partial^n f}{\partial x^n}(0).$$

A fundamental result in the theory of regular functions, that relates slice regularity and classical holomorphy, is the following, [13]:

Lemma 1 (Splitting Lemma) *If f is a regular function on $B = B(0, R)$, then for every $I \in \mathbb{S}$ and for every $J \in \mathbb{S}$, J orthogonal to I, there exist two holomorphic functions $F, G : B_I \to L_I$, such that for every $z = x + yI \in B_I$, it holds*

$$f_I(z) = F(z) + G(z)J.$$

The following version of the Identity Principle is one of the first consequences, (as shown in [13]):

Theorem 2 (Identity Principle) *Let f be a regular function on $B = B(0, R)$. Denote by Z_f the zero set of f, $Z_f = \{q \in B \mid f(q) = 0\}$. If there exists $I \in \mathbb{S}$ such that $B_I \cap Z_f$ has an accumulation point in B_I, then f vanishes identically on B.*

Another useful result is the following (see [2, 3]):

Theorem 3 (Representation Formula) *Let f be a regular function on $B = B(0, R)$ and let $J \in \mathbb{S}$. Then, for all $x + yI \in B$, the following equality holds*

$$f(x + yI) = \frac{1}{2}\big[f(x + yJ) + f(x - yJ)\big] + I\frac{1}{2}\big[J\big[f(x - yJ) - f(x + yJ)\big]\big].$$

In particular for each sphere of the form $x + y\mathbb{S}$ contained in B, there exist $b, c \in \mathbb{H}$ such that $f(x + yI) = b + Ic$ for all $I \in \mathbb{S}$.

Thanks to this result, it is possible to recover the values of a function on more general domains than open balls centred at the origin, from its values on a single slice L_I. This yields an extension theorem (see [2, 3]) that in the special case of functions that are regular on $B(0, R)$ can be obtained by means of their power series expansion.

Remark 1 If f_I is a holomorphic function on a disc $B_I = B(0, R) \cap L_I$ and its power series expansion is

$$f_I(z) = \sum_{n=0}^{\infty} z^n a_n, \quad \text{with } \{a_n\}_{n \in \mathbb{N}} \subset \mathbb{H},$$

then the unique regular extension of f_I to the whole ball $B(0, R)$ is the function defined as

$$\mathrm{ext}(f_I)(q) = \sum_{n=0}^{\infty} q^n a_n.$$

The uniqueness is guaranteed by the Identity Principle 2.

If we multiply pointwise two regular functions in general we will not obtain a regular function. To guarantee the regularity of the product we need to introduce a new multiplication operation, the $*$-product. On open balls centred at the origin we can define the $*$-product of two regular functions by means of their power series expansions, see [9].

Definition 3 Let $f, g : B = B(0, R) \to \mathbb{H}$ be regular functions and let

$$f(q) = \sum_{n=0}^{\infty} q^n a_n, \qquad g(q) = \sum_{n=0}^{\infty} q^n b_n$$

be their series expansions. The *regular product* (or $*$*-product*) of f and g is the function defined as

$$f * g(q) = \sum_{n=0}^{\infty} q^n \left(\sum_{k=0}^{n} a_k b_{n-k} \right),$$

regular on B.

Notice that the $*$-product is associative but generally it is not commutative. Its connection with the usual pointwise product is stated by the following result.

Proposition 1 *Let $f(q)$ and $g(q)$ be regular functions on $B = B(0, R)$. Then, for all $q \in B$,*

$$f * g(q) = \begin{cases} f(q)g(f(q)^{-1}qf(q)) & \text{if } f(q) \neq 0 \\ 0 & \text{if } f(q) = 0. \end{cases} \tag{1}$$

We remark that if $q = x + yI$ (and if $f(q) \neq 0$), then $f(q)^{-1}qf(q)$ has the same modulus and same real part as q. Therefore $f(q)^{-1}qf(q)$ lies in the same 2-sphere $x + y\mathbb{S}$ as q. We obtain then that a zero $x_0 + y_0 I$ of the function g is not necessarily a zero of $f * g$, but an element on the same sphere $x_0 + y_0\mathbb{S}$ does. In particular a real zero of g is still a zero of $f * g$. To present a characterization of the structure of the zero set of a regular function f we need to introduce the following functions.

Definition 4 Let $f(q) = \sum_{n=0}^{\infty} q^n a_n$ be a regular function on $B = B(0, R)$. We define the *regular conjugate* of f as

$$f^c(q) = \sum_{n=0}^{\infty} q^n \overline{a_n},$$

and the *symmetrization* of f as

$$f^s(q) = f * f^c(q) = f^c * f(q) = \sum_{n=0}^{\infty} q^n \left(\sum_{k=0}^{n} a_k \overline{a_{n-k}} \right).$$

Both f^c and f^s are regular functions on B.

Remark 2 Let $f(q) = \sum_{n=0}^{\infty} q^n a_n$ be regular on $B = B(0, R)$ and let $I \in \mathbb{S}$. Consider the splitting of f on L_I, $f_I(z) = F(z) + G(z)J$ with $J \in \mathbb{S}$, J orthogonal to I and F, G holomorphic functions on L_I. In terms of power series, if $F(z) = \sum_{n=0}^{\infty} z^n \alpha_n$ and $G(z) = \sum_{n=0}^{\infty} z^n \beta_n$ we have

$$f_I(z) = \sum_{n=0}^{\infty} z^n a_n = \sum_{n=0}^{\infty} z^n \alpha_n + \sum_{n=0}^{\infty} z^n \beta_n J = \sum_{n=0}^{\infty} z^n (\alpha_n + \beta_n J).$$

Hence the regular conjugate has splitting

$$f_I^c(z) = \sum_{n=0}^{\infty} z^n \overline{(\alpha_n + \beta_n J)} = \sum_{n=0}^{\infty} z^n \overline{\alpha_n} - \sum_{n=0}^{\infty} z^n \beta_n J.$$

That is

$$f_I^c(z) = \overline{F(\overline{z})} - G(z)J.$$

The function f^s is slice preserving (see [2]), i.e. $f^s(L_I) \subset L_I$ for every $I \in \mathbb{S}$. Thanks to this property it is possible to prove (see for instance [9]) that the zero set of a regular function that does not vanishes identically, consists of isolated points or isolated 2-spheres of the form $x + y\mathbb{S}$ with $x, y \in \mathbb{R}$, $y \neq 0$.

A calculation shows that the slice derivative satisfies the Leibniz rule with respect to the $*$-product.

Proposition 2 (Leibniz rule) *Let f and g be regular functions on $B = B(0, R)$. Then*

$$\partial_C(f * g)(q) = \partial_C f * g(q) + f * \partial_C g(q)$$

for every $q \in B$.

A basic result in analogy with the complex case, is the following (see [10]).

Theorem 4 (Maximum Modulus Principle) *Let $f : B = B(0, R) \to \mathbb{H}$ be a regular function. If $|f|$ has a local maximum in B, then f is constant in B.*

In [14] it is proved that we can estimate the maximum modulus of a function with its maximum modulus on each slice.

Proposition 3 *Let f be a regular function on $B = B(0, R)$. If there exist an imaginary unit $I \in \mathbb{S}$ and a real number $m \in (0, +\infty)$ such that*

$$f_I(B_I) \subset B(0, m),$$

then

$$f(B) \subset B(0, 2m).$$

In particular, if we set $m = \sup_{z \in B_I} |f(z)|$, then

$$\sup_{q \in B} \big|f(q)\big| \le 2m. \tag{2}$$

3 Uniform Norm and Regular Conjugation

This section is devoted to prove that the uniform norm on an open ball centred at the origin is the same for a regular function and for its regular conjugate.

Proposition 4 *Let c be in $\mathbb{H}$. Then the sets $\{cI \mid I \in \mathbb{S}\}$ and $\{Ic \mid I \in \mathbb{S}\}$ do coincide.*

Proof Let $c = a + bJ$ with $a, b \in \mathbb{R}$ and $J \in \mathbb{S}$. Let us fix $I \in \mathbb{S}$. We want to find an element L of $\mathbb{S}$ such that $cI = Lc$, that is such that

$$aI + bJI = aL + bLJ. \tag{3}$$

Let us denote by $\langle \, , \rangle$ the usual scalar product and by $\times$ the vector product. Recalling that for all imaginary units $I, J \in \mathbb{S}$ the following multiplication rule holds, see [13],

$$IJ = -\langle I, J\rangle + I \times J,$$

we can write Eq. (3) as

$$aI + b\big(-\langle J, I\rangle + J \times I\big) = aL + b\big(-\langle L, J\rangle + L \times J\big).$$

If we complete J to a orthonormal basis $1, J, K, JK$, of $\mathbb{H}$ over $\mathbb{R}$, then we can decompose

$$I = i_1 J + i_2 K + i_3 JK \quad \text{and} \quad L = l_1 J + l_2 K + l_3 JK,$$

obtaining that

$$J \times I = -i_3 K + i_2 JK \quad \text{and} \quad L \times J = l_3 K - l_2 JK.$$

Hence we need an imaginary unit L such that

$$\begin{aligned} &a(i_1 J + i_2 K + i_3 JK) + b(-i_1 + -i_3 K + i_2 JK) \\ &\quad = a(l_1 J + l_2 K + l_3 JK) + b(-l_1 + l_3 K - l_2 JK). \end{aligned}$$

Considering the different components along $1, J, K, JK$, we get that L has to satisfy the following linear system

$$\begin{cases} bi_1 = bl_1 \\ ai_1 = al_1 \\ ai_2 - bi_3 = al_2 + bl_3 \\ bi_2 + ai_3 = -bl_2 + al_3 \end{cases} \tag{4}$$

that has a unique solution (l_1, l_2, l_3) determining L. □

Proposition 5 *Let f be a regular function on $B = B(0, R)$. For any sphere of the form $x + y\mathbb{S}$ contained in B the following equalities hold:*

$$\inf_{I\in\mathbb{S}}\left|f(x+yI)\right| = \inf_{I\in\mathbb{S}}\left|f^c(x+yI)\right| \quad \text{and} \quad \sup_{I\in\mathbb{S}}\left|f(x+yI)\right| = \sup_{I\in\mathbb{S}}\left|f^c(x+yI)\right|.$$

Proof Let $q = x + yI$ be an element of B. Theorem 3 yields that f is affine on the sphere $x + y\mathbb{S}$ and there exist $b, c \in \mathbb{H}$ such that $f(x + yI) = b + Ic$ for all $I \in \mathbb{S}$. We want to compare now the value of f with the one attained by f^c by means of their power series expansions. If f has power series expansion $f(q) = \sum_{n\geq 0} q^n a_n$ and we set $w = x + yJ$, then by Theorem 3 we get

$$\begin{aligned} f(q) = f(x+yI) &= \frac{1}{2}\Big(\sum_{n\geq 0} w^n a_n + \sum_{n\geq 0} \overline{w^n} a_n\Big) + \frac{IJ}{2}\Big(\sum_{n\geq 0} \overline{w^n} a_n - \sum_{n\geq 0} w^n a_n\Big) \\ &= \frac{1}{2}\sum_{n\geq 0}(w^n + \overline{w^n})a_n + \frac{IJ}{2}\sum_{n\geq 0}(\overline{w^n} - w^n)a_n \\ &= \frac{1}{2}\sum_{n\geq 0} 2\,\mathrm{Re}(w^n)a_n + \frac{IJ}{2}\sum_{n\geq 0} -2\,\mathrm{Im}(w^n)a_n \\ &= \sum_{n\geq 0}\mathrm{Re}(w^n)a_n + I\sum_{n\geq 0}\left|\mathrm{Im}(w^n)\right|a_n. \end{aligned} \tag{5}$$

Hence the constants b and c are

$$b = \sum_{n\geq 0}\mathrm{Re}(w^n)a_n \quad \text{and} \quad c = \sum_{n\geq 0}\left|\mathrm{Im}(w^n)\right|a_n.$$

Since the power series expansion of f^c is $f^c(q) = \sum_{n\geq 0} q^n\overline{a_n}$, we obtain that for all $I \in \mathbb{S}$

$$f^c(x+yI) = \sum_{n\geq 0}\mathrm{Re}(w^n)\overline{a_n} + I\sum_{n\geq 0}\left|\mathrm{Im}(w^n)\right|\overline{a_n}.$$

Notice that $\mathrm{Re}(w^n)$ and $|\,\mathrm{Im}(w^n)| \in \mathbb{R}$ for all $n \geq 0$, then, in terms of b and c, we can write

$$f^c(x+yI) = \overline{b} + I\overline{c}$$

for all I in $\mathbb{S}$. Hence

$$\sup_{I\in\mathbb{S}}\left|f^c(x+yI)\right| = \sup_{I\in\mathbb{S}}|\overline{b} + I\overline{c}| = \sup_{I\in\mathbb{S}}|\overline{\overline{b} + I\overline{c}}| = \sup_{I\in\mathbb{S}}|b - cI| = \sup_{I\in\mathbb{S}}|b + cI|.$$

By Proposition 4 we obtain that

$$\sup_{I\in\mathbb{S}}\left|f^c(x+yI)\right| = \sup_{I\in\mathbb{S}}|b+cI| = \sup_{I\in\mathbb{S}}|b+Ic| = \sup_{I\in\mathbb{S}}\left|f(x+yI)\right|.$$

Exactly the same arguments hold for the infimum, so we can conclude also that

$$\inf_{I\in\mathbb{S}}\left|f^c(x+yI)\right| = \inf_{I\in\mathbb{S}}\left|f(x+yI)\right|. \qquad \square$$

Corollary 1 *Let f be a regular function on $B = B(0, R)$. Then*

$$\sup_{q\in B}\left|f(q)\right| = \sup_{q\in B}\left|f^c(q)\right| \quad \textit{and} \quad \inf_{q\in B}\left|f(q)\right| = \inf_{q\in B}\left|f^c(q)\right|.$$

Proof If S is the subset of $\mathbb{R}^2$ defined as

$$S = \left\{(x, y) \in \mathbb{R}^2 \mid y \geq 0,\ x^2 + y^2 \leq R^2\right\},$$

then we can cover the entire ball B with spheres of the form $x + y\mathbb{S}$ as

$$B = \bigcup_{(x,y)\in S} x + y\mathbb{S} = \bigcup_{(x,y)\in S}\bigcup_{I\in\mathbb{S}} x + yI.$$

By Proposition 5 we get

$$\begin{aligned}\sup_{q\in B}\left|f(q)\right| &= \sup_{(x,y)\in S}\sup_{I\in\mathbb{S}}\left|f(x+yI)\right| \\ &= \sup_{(x,y)\in S}\sup_{I\in\mathbb{S}}\left|f^c(x+yI)\right| = \sup_{q\in B}\left|f^c(q)\right|\end{aligned}$$

and the same holds for the infimum, hence we have also

$$\inf_{q\in B}\left|f(q)\right| = \inf_{q\in B}\left|f^c(q)\right|. \qquad \square$$

4 A Norm for a Mean Value Theorem

The main technical tool to prove an analog of the Bloch-Landau Theorem for regular functions is stated in terms of a new norm. This norm is equivalent to the uniform norm, on the space of functions that are regular on an open ball B centred at the origin, $B = B(0, R)$. The motivation to introduce this norm relies upon one of its properties, stated at the end of the section, which is useful to prove a mean value theorem.

Let $f : B \to \mathbb{H}$ be a regular function. Take $I, J \in \mathbb{S}$, I orthogonal to J and, according to the Splitting Lemma 1, let $F, G : B_I \to L_I$ be the holomorphic functions such that the restriction of f to $B_I = B \cap L_I$ is

$$f_I(z) = F(z) + G(z)J.$$

Let Ω be a subset of the ball B, and let $\|\cdot\|_\Omega$ denote the uniform norm on $\Omega \subseteq B$,

$$\|\cdot\|_\Omega = \sup_\Omega|\cdot|.$$

For any $I \in \mathbb{S}$, we will indicate with $\|\cdot\|_I$ the function

$$\|\cdot\|_I : \{f : B \to \mathbb{H} \mid f \text{ is regular}\} \to [0, +\infty)$$

defined by

$$\|f\|_I^2 = \|F\|_{B_I}^2 + \|G\|_{B_I}^2.$$

Remark 3 For all $I \in \mathbb{S}$, the function $\|\cdot\|_I$ does not depend on J; in fact, if we choose another imaginary unit $K \in \mathbb{S}$, orthogonal to I, then the splitting of f on L_I is

$$f_I(z) = \tilde{F}(z) + \tilde{G}(z)K,$$

where $\tilde{F}(z) = F(z)$, because I and K are orthogonal, and hence $\tilde{G}(z) = G(z)JK^{-1}$. Then

$$\left|\tilde{G}(z)\right| = \left|G(z)JK^{-1}\right| = \left|G(z)\right|$$

for all z in B_I, and hence $\|f\|_I$ does not change.

Consider now the function

$$\|\cdot\| : \{f : B \to \mathbb{H} \mid f \text{ is regular}\} \to [0, +\infty)$$

defined by

$$\|f\| = \sup_{I \in \mathbb{S}} \|f\|_I.$$

Proposition 6 *The function $\|\cdot\|$ is a norm on the real vector space $\mathscr{B} = \{f : B \to \mathbb{H} \mid f$ is regular$\}$.*

Proof Let $f \in \mathscr{B}$, $I \in \mathbb{S}$, and take $J \in \mathbb{S}$ orthogonal to I. Let F, and G be the holomorphic functions on L_I, such that $f_I(z) = F(z) + G(z)J$ for all $z \in B_I$. Then:

- $\|f\| = 0$ if and only if, for all $I \in \mathbb{S}$,

$$0 = \|f\|_I^2 = \|F\|_{B_I}^2 + \|G\|_{B_I}^2,$$

  and hence if and only if $F = G = 0$. By the Identity Principle 2 we can conclude that $\|f\| = 0$ if and only if $f = 0$.
- Let $c \in \mathbb{R}$. Then the splitting of fc on L_I is $(fc)_I(z) = F(z)c + G(z)cJ$. Hence, using the homogeneity of the uniform norm, we have

$$\begin{aligned}
\|fc\|^2 &= \sup_{I \in \mathbb{S}} \|fc\|_I^2 \\
&= \sup_{I \in \mathbb{S}} \left(\|F\|_{B_I}^2 |c|^2 + \|G\|_{B_I}^2 |c|^2\right) \\
&= \sup_{I \in \mathbb{S}} \left(\|F\|_{B_I}^2 + \|G\|_{B_I}^2\right) |c|^2 = |c|^2 \|f\|^2.
\end{aligned}$$

- If F_j, G_j are the splitting functions of regular functions f_j with respect to I and J for $j = 1, 2$, then

$$\begin{aligned}
&(\|f_1\|_I + \|f_2\|_I)^2 \\
&= \left(\sqrt{\|F_1\|^2_{B_I} + \|G_1\|^2_{B_I}} + \sqrt{\|F_2\|^2_{B_I} + \|G_2\|^2_{B_I}}\right)^2 \\
&= \|F_1\|^2_{B_I} + \|G_1\|^2_{B_I} + \|F_2\|^2_{B_I} + \|G_2\|^2_{B_I} \\
&\quad + 2\sqrt{(\|F_1\|^2_{B_I} + \|G_1\|^2_{B_I})(\|F_2\|^2_{B_I} + \|G_2\|^2_{B_I})},
\end{aligned} \tag{6}$$

and

$$\begin{aligned}
\|f_1 + f_2\|^2_I &= \|F_1 + F_2\|^2_{B_I} + \|G_1 + G_2\|^2_{B_I} \\
&= \left\|(F_1 + F_2)^2\right\|_{B_I} + \left\|(G_1 + G_2)^2\right\|_{B_I} \\
&= \left\|F_1^2 + F_2^2 + 2F_1F_2\right\|_{B_I} + \left\|G_1^2 + G_2^2 + 2G_1G_2\right\|_{B_I} \\
&\le \left\|F_1^2\right\|_{B_I} + \left\|F_2^2\right\|_{B_I} + 2\|F_1\|_{B_I}\|F_2\|_{B_I} \\
&\quad + \left\|G_1^2\right\|_{B_I} + \left\|G_2^2\right\|_{B_I} + 2\|G_1\|_{B_I}\|G_2\|_{B_I}.
\end{aligned} \tag{7}$$

The last quantity in Eq. (7) is less or equal than $(\|f_1\|_I + \|f_2\|_I)^2$ if and only if

$$\begin{aligned}
&\|F_1\|_{B_I}\|F_2\|_{B_I} + \|G_1\|_{B_I}\|G_2\|_{B_I} \\
&\le \sqrt{(\|F_1\|^2_{B_I} + \|G_1\|^2_{B_I})(\|F_2\|^2_{B_I} + \|G_2\|^2_{B_I})},
\end{aligned}$$

that is, if and only if

$$\begin{aligned}
&\left\langle (\|F_1\|_{B_I}, \|G_1\|_{B_I}), (\|F_2\|_{B_I}, \|G_2\|_{B_I}) \right\rangle \\
&\le \sqrt{\|F_1\|^2_{B_I} + \|G_1\|^2_{B_I}}\sqrt{\|F_2\|^2_{B_I} + \|G_2\|^2_{B_I}},
\end{aligned}$$

that holds thanks to Cauchy-Schwarz inequality for the scalar product on $\mathbb{R}^2$.

Therefore the function $\|\cdot\|$ is a norm. □

Let us now show that the norms $\|\cdot\|$ and $\|\cdot\|_B$, defined on $\mathscr{B}$, are equivalent.

Proposition 7 *Let $f : B = B(0, R) \to \mathbb{H}$ be a regular function. Then*

$$\frac{\sqrt{2}}{2}\|f\| \le \|f\|_B \le \|f\|.$$

Proof Let $I, J \in \mathbb{S}$, I orthogonal to J, and let F, G be holomorphic functions on L_I, such that $f_I(z) = F(z) + G(z)J$ for all $z \in B_I$. Then we have

$$\begin{aligned}
\|f\|^2_B &= \sup_{q\in B}\left|f(q)\right|^2 = \sup_{I\in\mathbb{S}}\sup_{z\in B_I}\left|f_I(z)\right|^2 \\
&= \sup_{I\in\mathbb{S}}\sup_{z\in B_I}\left|F(z) + G(z)J\right|^2 = \sup_{I\in\mathbb{S}}\sup_{z\in B_I}\left(\left|F(z)\right|^2 + \left|G(z)\right|^2\right)
\end{aligned}$$

$$\leq \sup_{I\in\mathbb{S}}\Big(\sup_{z\in B_I}|F(z)|^2+\sup_{z\in B_I}|G(z)|^2\Big)$$
$$=\sup_{I\in\mathbb{S}}\big(\|F\|_{B_I}^2+\|G\|_{B_I}^2\big)=\|f\|^2.$$

Conversely

$$\|f\|^2=\sup_{I\in\mathbb{S}}\|f\|_I^2=\sup_{I\in\mathbb{S}}\big(\|F\|_{B_I}^2+\|G\|_{B_I}^2\big)$$
$$\leq \sup_{I\in\mathbb{S}}\big(\|f_I\|_{B_I}^2+\|f_I\|_{B_I}^2\big)=2\|f\|_B^2.$$ □

As we anticipated at the beginning of this section, in terms of the norm $\|\cdot\|$ we can state (and prove) a mean value theorem.

Theorem 5 *Let f be a regular function on $B=B(0,R)$ such that $f(0)=0$. Then*

$$|q^{-1}f(q)|\leq\|\partial_C f\|$$

for all $q\in B$.

Proof Let $I\in\mathbb{S}$ and take $J\in\mathbb{S}$ orthogonal to I. By the Splitting Lemma 1 there exist $F,G:B_I\to L_I$ holomorphic functions, such that $f_I(z)=F(z)+G(z)J$ for all $z\in B_I$. By the Fundamental Theorem of Calculus for holomorphic functions (see for instance Theorem 3.2.1 in book [20]), we get that

$$|z^{-1}F(z)|\leq\|F'\|_{B_I}\quad\text{and}\quad|z^{-1}G(z)|\leq\|G'\|_{B_I}$$

for all $z\in B_I$. Hence

$$\begin{aligned}|z^{-1}f(z)|^2&=|z^{-1}F(z)+z^{-1}G(z)J|^2=|z^{-1}F(z)|^2+|z^{-1}G(z)|^2\\&\leq\|F'\|_{B_I}^2+\|G'\|_{B_I}^2=\|\partial_C f\|_I^2\\&\leq\sup_{I\in\mathbb{S}}\|\partial_C f\|_I^2=\|\partial_C f\|^2.\end{aligned}\tag{8}$$

Since $\|\partial_C f\|$ does not depend on I, we have that inequality (8) holds for every $q\in B$. □

Remark 4 As a consequence of Theorem 5 and of the Maximum Modulus Principle 4, we get also that if f is regular on $B(0,R)$, then for all $r\in(0,R)$

$$\sup_{q\in B(0,r)}|f(q)|\leq r\|\partial_C f\|.\tag{9}$$

We conclude this section showing that, as the uniform norm, the norm $\|\cdot\|$ satisfies the following

Proposition 8 *Let f be a regular function on $B=B(0,R)$. Then $\|f\|=\|f^c\|$.*

Proof Let $I \in \mathbb{S}$. By Remark 2, if the splitting of f_I is $f_I(z) = F(z) + G(z)J$ for all $z \in B_I$, then the regular conjugate of f splits as

$$f_I^c(z) = \overline{F(\overline{z})} - G(z)J.$$

Since the (complex) conjugation is a bijection of B_I and the modulus $|\overline{F(z)}|$ is equal to $|F(z)|$ for all $z \in B_I$, we get that

$$\|f^c\|_I^2 = \|\overline{F(\overline{z})}\|_{B_I}^2 + \|-G\|_{B_I}^2 = \|F\|_{B_I}^2 + \|G\|_{B_I}^2 = \|f\|_I^2. \tag{10}$$

Since equality (10) holds for every I in $\mathbb{S}$, we can conclude. □

5 The Bloch-Landau Type Theorem

For $\rho > 0$, let us denote by $\mathcal{O}(\rho)$ the open set

$$\mathcal{O}(\rho) = \{q \in \mathbb{H} \mid |q|^3 < \rho|\mathrm{Re}(q)|^2\}.$$

Notice that the intersection of $\mathcal{O}(\rho)$ with a slice $L_I = \{x + yI \mid x, y \in \mathbb{R}\}$ is the interior of an eight shaped curve with equation

$$(x^2 + y^2)^{\frac{3}{2}} = \rho x^2.$$

We remark that this curve always contains two discs with positive radius depending just on ρ. For example, we can take the discs centred in $(\frac{\rho}{2}, 0)$ and in $(-\frac{\rho}{2}, 0)$, of radius $\frac{37}{256}\rho^2$. Therefore, the open set $\mathcal{O}(\rho)$ always contains two open balls of radius at least $\frac{37}{256}\rho^2$.

Lemma 2 *Let $f : B = B(0, R) \to \mathbb{H}$ be a (non constant) regular function such that $f(0) = 0$ and that $\partial_C f(0) \in \mathbb{R}$. Then the image of B under f contains an open set of the form $\mathcal{O}(\rho)$ where*

$$\rho = \frac{R|\partial_C f(0)|^2}{4\|\partial_C f\|}.$$

Proof Since $f(0) = 0$, if $\partial_C f(0) = 0$ there is nothing to prove. Suppose then that $\partial_C f(0) \neq 0$. Consider a point c outside the image of B under f, then $c \neq 0$. We want to show that c does not belong to $\mathcal{O}(\rho)$.

For all $q \in B$, define $g(q)$ to be

$$g(q) = (1 - f(q)c^{-1})^s.$$

The function g is regular on B and we can estimate its modulus in the following manner: let $\tau(q)$ be the transformation defined by

$$\tau(q) = (1 - f(q)c^{-1})^{-1} q (1 - f(q)c^{-1}).$$

Then, according to Proposition 1 and recalling that $|\tau(q)| = |q|$ for all q, we can write

$$\begin{aligned}|g(q)| &= \left|\left(1-f(q)c^{-1}\right)*\left(1-f(q)c^{-1}\right)^c\right| \\ &= \left|\left(1-f(q)c^{-1}\right)\right|\left|\left(1-f(\tau(q))c^{-1}\right)^c\right| \\ &\leq \sup_{|q|<R}\left|\left(1-f(q)c^{-1}\right)\right| \sup_{|q|<R}\left|\left(1-f(q)c^{-1}\right)^c\right|.\end{aligned}$$

Hence, using Proposition 5,

$$|g(q)| \leq \left(\sup_{|q|<R}\left|\left(1-f(q)c^{-1}\right)\right|\right)^2,$$

that is

$$|g(q)|^{\frac{1}{2}} \leq \sup_{|q|<R}\left|\left(1-f(q)c^{-1}\right)\right|.$$

By the properties of the uniform norm and by Remark 4 we get then

$$|g(q)|^{\frac{1}{2}} \leq 1+\sup_{|q|<R}\left|f(q)c^{-1}\right| \leq 1+|c|^{-1}\sup_{|q|<R}\left|f(q)\right| \leq 1+|c|^{-1}\|\partial_C f\|R. \quad (11)$$

The next step is to estimate from below the quantity $|g(q)|^{\frac{1}{2}}$. Notice that g is slice preserving, since it is the symmetrization of a regular function. Moreover $g(q)$ is never zero. For all I in $\mathbb{S}$ the map $z \mapsto z^4$ from $L_I \setminus \{0\} \to L_I \setminus \{0\}$ is a covering map. Since B_I is simply connected, we can lift the function g obtaining a holomorphic function $\Psi_I : B_I \to L_I \setminus \{0\}$ such that

$$\Psi_I(z)^4 = \left(1-f(z)c^{-1}\right)^s$$

and $\Psi_I(0) = 1$ (since $g(0) = 1$). Let Ψ be the (unique) extension to B of Ψ_I. Now $\Psi(0) = 1$,

$$\Psi(q)^4 = \left(1-f(q)c^{-1}\right)^s = g(q)$$

for all $q \in B$ and in particular

$$|\Psi(q)|^2 = |g(q)|^{\frac{1}{2}} \quad \text{for all } q \in B. \quad (12)$$

We want to use the power series expansion of Ψ to find a lower bound of $|g|^{\frac{1}{2}}$. In particular we need to compute its slice derivative. Using the Leibniz rule 2 we can calculate

$$\begin{aligned}\partial_C g(q) &= \partial_C\left[\left(1-f(q)c^{-1}\right)^s\right] = \partial_C\left[\left(1-f(q)c^{-1}\right)*\left(1-f(q)c^{-1}\right)^c\right] \\ &= -\partial_C f(q)c^{-1}*\left(1-\left(f^c(q)c^{-1}\right)^c\right)-\left(1-f(q)c^{-1}\right)*\partial_C\left(f(q)c^{-1}\right)^c.\end{aligned}$$

Since $q = 0$ is a real zero of f (and hence of f^c), and since the operators of slice differentiation and regular conjugation do commute, if the power series expansion of f is $f(q) = \sum_{n=0}^{\infty} q^n a_n$, we obtain that

$$\left(\partial_C f(q)c^{-1}\right)^c = \sum_{n=1}^{\infty} q^{n-1}\overline{n a_n c^{-1}}$$

and hence that

$$\partial_C g(0) = -\partial_C f(0)c^{-1} - \overline{\partial_C f(0)c^{-1}} = -2\,\mathrm{Re}\big(\partial_C f(0)c^{-1}\big).$$

Moreover, since g (and Ψ) is slice preserving, $g(0) = 1$, and $\partial_C f(0)$ is real, we have

$$\partial_C \Psi(0) = \frac{1}{4} g(0)^{\frac{1}{4}-1} \partial_C g(0) = -\frac{1}{2}\,\mathrm{Re}\big(\partial_C f(0)c^{-1}\big) = -\frac{1}{2}\partial_C f(0)\,\mathrm{Re}\big(c^{-1}\big). \quad (13)$$

Let us set

$$M := 1 + |c|^{-1} \|\partial_C f\| R.$$

Fix $r \in (0, R)$ and $I \in \mathbb{S}$. Let $q \in \partial B_I(0, r)$, $q = re^{I\theta}$ for some $\theta \in [0, 2\pi)$. By Eqs. (11) and (12) we obtain that for every $\theta \in [0, 2\pi)$

$$M \geq \big|\Psi\big(re^{I\theta}\big)\big|^2. \quad (14)$$

Using the series expansion of Ψ we can write

$$\begin{aligned} \big|\Psi\big(re^{I\theta}\big)\big|^2 &= \overline{\Psi\big(re^{I\theta}\big)}\Psi\big(re^{I\theta}\big) \\ &= \left(\sum_{m=0}^{\infty} \frac{1}{m!} r^m \overline{\Psi^{(m)}(0)} e^{-Im\theta}\right)\left(\sum_{n=0}^{\infty} \frac{1}{n!} r^n e^{In\theta} \Psi^{(n)}(0)\right) \\ &= \sum_{m,n=0}^{\infty} \frac{1}{m!n!} r^{m+n} \overline{\Psi^{(m)}(0)} e^{I(n-m)\theta} \Psi^{(n)}(0). \end{aligned} \quad (15)$$

If we integrate in θ, then we get

$$\frac{1}{2\pi}\int_0^{2\pi} \big|\Psi\big(re^{I\theta}\big)\big|^2 d\theta = \frac{1}{2\pi}\int_0^{2\pi} \sum_{m,n=0}^{\infty} \frac{1}{m!n!} r^{m+n} \overline{\Psi^{(m)}(0)} e^{I(n-m)\theta} \Psi^{(n)}(0)\, d\theta.$$

Thanks to the uniform convergence on compact sets of the series expansion, we can exchange the order of integration and summation. Then, since $\int_0^{2\pi} e^{Is\theta}\, d\theta = 0$ if $s \in \mathbb{Z}$, $s \neq 0$ and equals 2π otherwise, just the terms where $n = m$ survive and hence we get

$$\frac{1}{2\pi}\int_0^{2\pi} \big|\Psi\big(re^{I\theta}\big)\big|^2 d\theta = \sum_{m=0}^{\infty} \frac{r^{2m}}{(m!)^2} \overline{\Psi^{(m)}(0)} \Psi^{(m)}(0) = \sum_{m=0}^{\infty} \frac{r^{2m}}{(m!)^2} \big|\Psi^{(m)}(0)\big|^2.$$

By inequality (14) and since M is a constant, we obtain

$$M = \frac{1}{2\pi}\int_0^{2\pi} M\, d\theta \geq \sum_{m=0}^{\infty} \frac{r^{2m}}{(m!)^2} \big|\Psi^{(m)}(0)\big|^2.$$

Considering just the first two terms of the series expansion and using Eq. (13), we get

$$M \geq 1 + r^2 \big|\partial_C \Psi(0)\big|^2 = 1 + \frac{r^2 |\partial_C f(0)|^2 |\,\mathrm{Re}(c^{-1})|^2}{4} = 1 + \frac{r^2 |\partial_C f(0)|^2 |\,\mathrm{Re}(c)|^2}{4|c|^4}.$$

Recalling the expression of M we have then

$$1 + |c|^{-1}\|\partial_C f\|R \geq 1 + \frac{r^2|\partial_C f(0)|^2|\operatorname{Re}(c)|^2}{4|c|^4}, \tag{16}$$

that is

$$|c|^3 \geq \frac{r^2|\partial_C f(0)|^2|\operatorname{Re}(c)|^2}{4\|\partial_C f\|R}$$

for all $r \in (0, R)$. Hence, if we take the limit as r approaches R, we obtain that if a point c is outside the image $f(B)$, then it satisfies the following inequality

$$|c|^3 \geq \frac{R|\partial_C f(0)|^2|\operatorname{Re}(c)|^2}{4\|\partial_C f\|}.$$

That is equivalent to say that the image of B under f contains the open set

$$\mathscr{O}(\rho) = \{q \in \mathbb{H} \mid |q|^3 < \rho|\operatorname{Re}(q)|^2\},$$

where

$$\rho = \frac{R|\partial_C f(0)|^2}{4\|\partial_C f\|}. \qquad \square$$

Let $w \in \mathbb{H}$, and let τ_w be the translation $q \mapsto q + w$. The composition of a regular function with τ_w is not regular in general. For our purposes we need to define a new notion of composition in this special case. Notice that if we restrict both functions to the slice L_I containing w, then we can compose them obtaining a holomorphic function.

Definition 5 Let f be a regular function on $B = B(0, R)$ and let $w = x + yI \in B$. We define the *regular translation* $\tilde{f}_w$ of f to be the unique regular extension of $f_I \circ (\tau_w)_I$,

$$\tilde{f}_w(q) = \operatorname{ext}\big(f_I \circ (\tau_w)_I\big)(q),$$

regular on $B(0, R - |w|)$.

In the proof of the main result, it will be useful the following

Proposition 9 *Let f be a regular function on $B = B(0, R)$ and let w_n be a convergent sequence in B, such that* $\lim_{n\to\infty} w_n = w \in B$. *Set*

$$m = \max\big\{\{|w_n|,\ n \in \mathbb{N}\} \cup \{|w|\}\big\}.$$

Then $\tilde{f}_{w_n}$ converges to $\tilde{f}_w$ uniformly on compact subsets of $B(0, R - m)$.

Proof First of all notice that the maximum m exists because of the convergence of the sequence $\{w_n\}$. Clearly the sequence τ_{w_n} converges (uniformly on compact sets) to τ_w. Moreover, if $w_n \notin \mathbb{R}$ frequently, then (up to a subsequence) we define

$$I_n = \frac{\operatorname{Im}(w_n)}{|\operatorname{Im}(w_n)|} \quad \text{and} \quad \lim_{n\to\infty} I_n = I.$$

If, instead, there exists a natural number n_0 such that w_n is real for all $n > n_0$ then we choose any $I \in \mathbb{S}$ and set $I_n = I$ for all $n > n_0$ in what follows. In both cases, $I \in \mathbb{S}$ is such that $w \in L_I$.

By Theorem 3 we can write for all $q = x + yJ \in B(0, R - m)$

$$\begin{aligned}
\tilde{f}_{w_n}(x+yJ) &= \frac{1}{2}\big[\tilde{f}_{w_n}(x+yI_n) + \tilde{f}_{w_n}(x-yI_n)\big] \\
&\quad + \frac{JI_n}{2}\big[\tilde{f}_{w_n}(x-yI_n) - \tilde{f}_{w_n}(x+yI_n)\big] \\
&= \frac{1}{2}\big[f_{I_n}(x+yI_n+w_n) + f_{I_n}(x-yI_n+w_n)\big] \\
&\quad + \frac{JI_n}{2}\big[f_{I_n}(x-yI_n+w_n) - f_{I_n}(x+yI_n+w_n)\big] \\
&= \frac{1}{2}\big[f(x+yI_n+w_n) + f(x-yI_n+w_n)\big] \\
&\quad + \frac{JI_n}{2}\big[f(x-yI_n+w_n) - f(x+yI_n+w_n)\big]. \qquad (17)
\end{aligned}$$

Since f is a continuous function, again by Theorem 3 we can conclude that (uniformly on compact sets)

$$\lim_{n\to\infty} \tilde{f}_{w_n}(q) = \tilde{f}_w(q)$$

for all $q \in B(0, R - m)$. □

In order to prove the Bloch-Landau type theorem we need a last step.

Proposition 10 *Let $f : B(0, R) \to \mathbb{H}$ be a regular function. Then $M : [0, R) \to \mathbb{R}$ defined as*

$$M(s) = \max_{|q|\le s} |f(q)|$$

is a continuous function.

Proof If f is constant the statement is trivially true. Let us suppose then that f is not constant. The Maximum Modulus Principle 4 yields that the function $M(s)$ is increasing and hence for any sequence s_n converging (from above or from below) to s there exists the limit $\lim_{n\to\infty} M(s_n)$. To show that the limit is equal to $M(s)$, consider first the sequence $\{s + \frac{1}{n}\}_{n\in\mathbb{N}}$. Since $B(0, s + \frac{1}{n})$ is relatively compact, we can find q_n such that $M(s + \frac{1}{n}) = |f(q_n)|$ for all $n \in \mathbb{N}$, and, up to subsequences, we can suppose that q_n converges to $q_0 \in \partial B(0, s)$. Therefore we have

$$M(s) \le \lim_{n\to\infty} M\left(s + \frac{1}{n}\right) = \lim_{n\to\infty} |f(q_n)| = |f(q_0)|,$$

where the last equality is due to the continuity of the function $|f(q)|$. Moreover, by definition of M we have that $|f(q_0)| \le M(s)$ and hence that

$$M(s) = \lim_{n\to\infty} M\left(s + \frac{1}{n}\right).$$

Now q_0 lies in the closure $\overline{B(0,s)}$, hence we can find a sequence whose term p_n has modulus $|p_n| = s - \frac{1}{n}$ for all $n \in \mathbb{N}$, such that $\lim_{n\to\infty} p_n = q_0$. Then

$$M(s) \geq \lim_{n\to\infty} M\left(s - \frac{1}{n}\right) \geq \lim_{n\to\infty} \left|f(p_n)\right| = \left|f(q_0)\right| = M(s).$$

Therefore we can conclude that M is continuous. □

Finally we have all the tools to prove the announced Bloch-Landau type theorem for regular functions. Let $\mathbb{B}$ be the unit open ball of $\mathbb{H}$, $\mathbb{B} = \{q \in \mathbb{H} \mid |q| < 1\}$.

Theorem 6 (Bloch-Landau type theorem) *Let $f : \mathbb{B} \to \mathbb{H}$ be a regular function such that $f(0) = 0$ and $\partial_C f(0) = 1$. Then there exists $u \in \mathbb{B}$ such that the image of the regular translation $\tilde{f}_u$ of f contains an open set obtained by means of a rotation and a translation of $\mathcal{O}(\rho)$, where the "radius" ρ is at least $\frac{1}{32\sqrt{2}}$.*

Proof Let us set $M(t)$ to be the function defined on $[0, 1)$ by

$$M(t) = \max_{|q|\leq t} \left|\partial_C f(q)\right|,$$

fix r in $(0, 1)$, and consider the function

$$\mu(s) = sM(r - s),$$

defined for $s \in [0, r]$. By Proposition 10, μ is a continuous function, $\mu(0) = 0$, $\mu(s) \geq 0$ for all $s \in [0, r]$, and $\mu(r) = r$. Set

$$R = \frac{1}{2} \min\{s \mid \mu(s) = r\},$$

then $0 < 2R \leq r$. Let $w \in \partial B(0, r - 2R)$ be such that $|\partial_C f(w)| = M(r - 2R)$, i.e. by definition of R, such that $|\partial_C f(w)| = \frac{r}{2R}$. Let us restrict our attention to the slice L_I containing w. Consider the function $\varphi_I : B(0, 2R) \cap L_I \to \mathbb{H}$, defined by

$$\varphi_I(z) = \big(f(z + w) - f(w)\big)\frac{\overline{\partial_C f(w)}}{|\partial_C f(w)|}.$$

The function φ_I is holomorphic on $B(0, 2R) \cap L_I$, because

$$|q + w| \leq |q| + |w| \leq 2R + (r - 2R) = r.$$

Let φ be the (unique) regular extension to the entire ball $B(0, 2R)$ of φ_I. Then $\varphi(0) = 0$ and $\partial_C\varphi(0) = |\partial_C f(w)| = \frac{r}{2R}$, hence φ satisfies the hypotheses of Lemma 2.

For $z \in B(0, R) \cap L_I$ we have that

$$\left|\partial_C\varphi_I(z)\right| = \left|\partial_C f(z + w)\right| \leq M\big(|z + w|\big) \leq M(r - R) = \frac{\mu(R)}{R}.$$

Since μ is continuous, $\mu(0) = 0$ and $\mu(r) = r$, then

$$\frac{\mu(R)}{R} < \frac{r}{R},$$

otherwise there would exists $s < 2R$ such that $\mu(s) = r$, a contradiction with the definition of R. Therefore

$$\partial_C \varphi_I\big(B(0, R) \cap L_I\big) \subset B\left(0, \frac{r}{R}\right).$$

Statement (2) of Proposition 3 implies then that

$$\partial_C \varphi\big(B(0, R)\big) \subset B\left(0, \frac{2r}{R}\right).$$

Considering the uniform norm we obtain

$$\|\partial_C \varphi\|_{B(0,R)} \leq \frac{2r}{R},$$

and hence, by Proposition 7,

$$\|\partial_C \varphi\| \leq \frac{2\sqrt{2}r}{R} \tag{18}$$

on $B(0, R)$. Lemma 2 yields then that $\varphi(B(0, R))$ contains an open set $\mathscr{O}(\rho)$ where

$$\rho = \frac{R|\partial_C \varphi(0)|^2}{4\|\partial_C \varphi\|} \geq \frac{R(\frac{r}{2R})^2}{4\frac{2\sqrt{2}r}{R}} = \frac{r}{32\sqrt{2}}.$$

Recalling the definition of φ, we get then

$$\tilde{f}_w\big(B(0, R)\big) - f(w) \supset \mathscr{O}\big(\rho(r)\big)\frac{\partial_C \varphi(0)}{|\partial_C \varphi(0)|},$$

that yields

$$f(w) + \mathscr{O}\big(\rho(r)\big)\frac{\partial_C \varphi(0)}{|\partial_C \varphi(0)|} \subset \tilde{f}_w\big(B(0, R)\big).$$

Therefore for all $r < 1$ there exist $R_r > 0$ and w_r, with modulus $|w_r| = r - 2R_r$, such that the image of $B(0, R_r)$ through $\tilde{f}_{w_r}$ contains the open set $f(w_r) + \mathscr{O}(\rho(r))\frac{\partial_C \varphi(0)}{|\partial_C \varphi(0)|}$. When r approaches 1, by compactness, we can find subsequences $\{R_n\}$ and $\{w_n\}$, converging respectively to $R_0 > 0$ and to $w_0 \in \mathbb{B}$ (in fact $R_0 = 0$ would imply that $\mu(0) = 1$ which is not). Thanks to Proposition 9 we have then that $\tilde{f}_{w_n}$ converges (uniformly on compact sets) to $\tilde{f}_{w_0}$, and hence we get that the image of $\tilde{f}_{w_0}$ contains the open set

$$\mathscr{O}(\rho)\frac{\partial_C \varphi(0)}{|\partial_C \varphi(0)|} + f(w_0)$$

whose "radius" is at least

$$\rho = \lim_{r\to 1} \rho(r) \geq \lim_{r\to 1} \frac{r}{32\sqrt{2}} = \frac{1}{32\sqrt{2}}.$$ □

It is easy to prove that if the regular translation $\tilde{f}_u$ that appears in the statement of Theorem 6 is a real translation (i.e. if u is real), then the universal set $\mathscr{O}(\frac{1}{32\sqrt{2}})$ is contained in $f(\mathbb{B})$.

It seems to us that, in general, there is not a universal open set directly contained in the image $f(\mathbb{B})$ of a (normalized) slice regular function f. And this might be connected with the fact that, as proved in [6], the Bloch-Landau Theorem does not hold in $\mathbb{C}^2$ (and $\mathbb{C}^2$ and $\mathbb{H}$ are strictly related).

In any case the authors plan to further investigate this fascinating subject in the near future.

References

1. Colombo, F., Sabadini, I., Sommen, F., Struppa, D.C.: Analysis of Dirac Systems and Computational Algebra. Birkhäuser, Basel (2004)
2. Colombo, F., Gentili, G., Sabadini, I., Struppa, D.: Extension results for slice regular functions of a quaternionic variable. Adv. Math. **222**, 1793–1808 (2009)
3. Colombo, F., Gentili, G., Sabadini, I.: A Cauchy kernel for slice regular functions. Ann. Glob. Anal. Geom. **37**, 361–378 (2010)
4. Colombo, F., Sabadini, I., Struppa, D.C.: Noncommutative Functional Calculus. Theory and Applications of Slice Hyperholomorphic Functions. Progress in Mathematics. Birkhäuser, Basel (2011)
5. Cullen, C.G.: An integral theorem for analytic intrinsic functions on quaternions. Duke Math. J. **32**, 139–148 (1965)
6. Duren, P., Rudin, W.: Distortion in several variables. Complex Var. Theory Appl. **5**, 323–326 (1986)
7. Fueter, R.: Die Funktionentheorie der Differentialgleichungen $\Delta u = 0$ und $\Delta\Delta u = 0$ mit vier reellen Variablen. Comment. Math. Helv. **7**, 307–330 (1934)
8. Fueter, R.: Über Hartogs'schen Satz. Comment. Math. Helv. **12**, 75–80 (1939)
9. Gentili, G., Stoppato, C.: Zeros of regular functions and polynomials of a quaternionic variable. Mich. Math. J. **56**, 655–667 (2008)
10. Gentili, G., Stoppato, C.: The open mapping theorem for regular quaternionic functions. Ann. Sc. Norm. Super. Pisa Cl. Sci. (5) **8**, 805–815 (2009)
11. Gentili, G., Stoppato, C.: Power series and analyticity over the quaternions. Math. Ann. **352**, 113–131 (2012)
12. Gentili, G., Struppa, D.C.: A new approach to Cullen-regular functions of a quaternionic variable. C. R. Acad. Sci. Paris **342**, 741–744 (2006)
13. Gentili, G., Struppa, D.C.: A new theory of regular function of a quaternionic variable. Adv. Math. **216**, 279–301 (2007)
14. Gentili, G., Vignozzi, I.: The Weierstrass factorization theorem for slice regular functions over the quaternions. Ann. Glob. Anal. Geom. **40**(4), 435–466 (2011)
15. Ghiloni, R., Perotti, A.: A new approach to slice regularity on real algebras. In: Sabadini, I., Sommen, F. (eds.) Hypercomplex Analysis and Its Applications. Trends in Mathematics, pp. 109–124. Birkhäuser, Basel (2011)
16. Ghiloni, R., Perotti, A.: Slice regular functions on real alternative algebras. Adv. Math. **226**(2), 1662–1691 (2011)
17. Gürlebeck, K., Morais, J.: Bloch's theorem in the context of quaternion analysis. arXiv: 1201.0530v1
18. Landau, E.: Über die Blochsche Konstante und zwei verwandte Weltkonstanten. Math. Z. **30**(1), 608–634 (1929)
19. Landau, E.: Darstellung und Begründung einiger neuerer Ergebnisse der Funktionentheorie. Springer, Berlin (1929)
20. Lang, S.: Complex Analysis. Graduate Texts in Mathematics, vol. 103. Springer, New York (1999)
21. Sudbery, A.: Quaternionic analysis. Math. Proc. Camb. Philos. Soc. **85**, 199–225 (1979)

Dirichlet-Type Problems for the Iterated Dirac Operator on the Unit Ball in Clifford Analysis

Min Ku, Uwe Kähler, and Paula Cerejeiras

Abstract We study a class of Dirichlet-type problems for null solutions to iterated Dirac operators on the unit ball of $\mathbf{R}^n$ with boundary data given by function in $\mathscr{L}_p$ $(1 < p < +\infty)$. Applying Almansi-type decomposition theorems for null solutions to iterated Dirac operators, our Dirichlet-type problems for null solutions to iterated Dirac operators is transferred to Dirichlet-type problems for monogenic functions or harmonic functions. By introducing shifted Euler operators and making use of Clifford-Cauchy transform, we get its unique solution and its integral representation.

1 Introduction

The Dirichlet problem for poly-harmonic and poly-analytic functions in the complex plane is fundamental for solving concrete problems in mathematical physics and engineering. A large number of investigations on the subject has been done (see, for instance, [1–5]) by making use of the intimate connection between harmonic functions and analytic functions in the complex plane. The corresponding higher-dimensional analogue, the Dirac equation, also plays very important role in problems in pure mathematics, mathematical physics, and engineering, such as the theory of specially relativistic quantum mechanics and quantum fields or in image analysis. The study of null solution to the Dirac equation is called Clifford analysis, i.e. the theory of monogenic functions (see e.g. [6–9]). It offers an intriguing generalization of complex analysis in the plane into higher dimensions and refines classic

M. Ku · U. Kähler (✉) · P. Cerejeiras
Center for Research and Development in Mathematics and Applications, Department of Mathematics, University of Aveiro, 3810-193 Aveiro, Portugal
e-mail: ukaehler@ua.pt

M. Ku
e-mail: kumin0844@gmail.com

M. Ku
e-mail: kumin0844@163.com

P. Cerejeiras
e-mail: pceres@ua.pt

G. Gentili et al. (eds.), *Advances in Hypercomplex Analysis*, Springer INdAM Series 1, DOI 10.1007/978-88-470-2445-8_5, © Springer-Verlag Italia 2013

multi-dimensional harmonic analysis due to the factorization of the Laplacian by the Dirac operator. In this setting, many results on various boundary value problems for monogenic functions, null solutions to iterated Dirac equations, and null solutions to polynomially generalized Cauchy-Riemann equations have widely been published, e.g. [10–19]. In [10–12], R. Delanghe, F. Sommen, T. Qian, M. Mitrea and others considered half Dirichlet problems on the unit ball, on the upper half space, and on Lipschitz surfaces of Euclidean space $\mathbf{R}^n$. In [13, 14], elliptic boundary value problems were studied. In [15–19], some Riemann-Hilbert boundary value problems for null solutions to the iterated Dirac equations and to polynomially generalized Cauchy-Riemann equations were discussed. More related results can be also seen in references [20–22]. In this setting, it seems natural to consider Dirichlet-type problems for null solutions to the iterated Dirac operator. However, as far as we know, it has not been discussed up to now. Based on ideas contained in [2, 10, 19], the purpose of this article is to think of Dirichlet-type problems for null solutions to iterated Dirac operator with given $\mathscr{L}_p$ $(1 < p < +\infty)$-boundary data on the unit ball of $\mathbf{R}^n$. Applying an Almansi-type decomposition theorem for iterated Dirac operator in Clifford analysis, we first transfer the problem into the corresponding Dirichlet-type problem for monogenic or harmonic functions. Then introducing shifted Euler operators and making full use of the Clifford-Cauchy transform, we get an integral representation of the unique solution to Dirichlet-type problems for null solutions to iterated Dirac operators with $\mathscr{L}_p$ $(1 < p < +\infty)$ boundary data on the unit ball of $\mathbf{R}^n$.

The paper is organized as follows. In Sect. 2, we recall some basic facts about Clifford analysis which will be needed in the sequel. In Sect. 3 we will present several technical lemmas. In Sect. 4 we will consider half Dirichlet problem on the unit ball $\mathbf{R}^n$ with boundary values given by $\mathscr{L}_p$ $(1 < p < +\infty)$-functions in Clifford analysis. Applying an Almansi-type decomposition for iterated Dirac operator in Clifford analysis, in virtue of shifted Euler operators and the Clifford-Cauchy transform we get the unique solution to it. In the last section, we further study the Dirichlet problem on the unit ball $\mathbf{R}^n$ for poly-harmonic functions, i.e. null solutions to iterated Dirac operator of even order, with boundary data given by $\mathscr{L}_p$ $(1 < p < +\infty)$-integral functions in Clifford analysis. Making full use of the Almansi decomposition theorem for poly-harmonic functions and Clifford-integral operators, we give the unique solution to it.

2 Preliminaries and Notations

In this section we recall some basic facts about Clifford analysis which will be needed in the sequel. More details can also be found in [6–8, 23–29].

Let $\{e_1, e_2, \ldots, e_n\}$ be an orthogonal basis of the Euclidean space $\mathbf{R}^n$, let $\mathbf{R}^n$ be endowed with a non-degenerate quadratic form of signature $(0, n)$ and let $\mathbf{R}_{0,n}$ be the 2^n-dimensional real Clifford algebra constructed over $\mathbf{R}^n$ with basis $\{e_{\mathscr{A}} : \mathscr{A} = \{h_1, \ldots, h_r\} \in \mathscr{PN},\ 0 \leq h_1 < h_r \leq n\}$, where $\mathscr{N}$ stands for the set $\{0, 1, 2, \ldots, n\}$ and $\mathscr{PN}$ denotes the family of all order-preserving subsets of $\mathscr{N}$. We denote by $e_\emptyset$ the identity element 1 and by $e_{h_1 \cdots h_r} = e_{\mathscr{A}}$ for $\mathscr{A} = \{h_1, \ldots, h_r\} \in \mathscr{PN}$. The product in $\mathbf{R}_{0,n}$ is defined by the rules: $e_i^2 = -1$ for $i = 1, 2, \ldots, n$ and $e_i e_j + e_j e_i = 0$ for $1 \le i < j \le n$. Thus the real Clifford algebra $\mathbf{R}_{0,n}$ is a real linear, associative, but non-commutative algebra. For arbitrary $a \in \mathbf{R}_{0,n}$ we have $a = \sum_{N(\mathscr{A})=k} a_{\mathscr{A}} e_{\mathscr{A}} = \sum_{N(\mathscr{A})=k} [a]_k$, $a_{\mathscr{A}} \in \mathbf{R}$, where $[a]_k = \sum_{N(\mathscr{A})=k} e_{\mathscr{A}} [a]_{\mathscr{A}}$ is the so-called k-vector part of a $(k = 0, 1, 2, \ldots, n)$. The Euclidean space $\mathbf{R}^n$ is embedded in $\mathbf{R}_{0,n}$ by identifying $(x_1, x_2, \ldots, x_n)$ with the Clifford vector x given by $x = \sum_{j=1}^n e_j x_j$. The conjugation in $\mathbf{R}_{0,n}$ is defined by $\bar{a} = \sum_{\mathscr{A}} a_{\mathscr{A}} \bar{e}_{\mathscr{A}}$, $\bar{e}_{\mathscr{A}} = (-1)^{\frac{k(k+1)}{2}} e_{\mathscr{A}}$, $N(\mathscr{A}) = k$, $a_{\mathscr{A}} \in \mathbf{R}$, and hence $\overline{ab} = \bar{b}\bar{a}$ for arbitrary $a, b \in \mathbf{R}_{0,n}$. Note that $x^2 = -\langle x, x \rangle = -|x|^2$. The complex Clifford algebra $\mathbf{C}_n = \mathbf{R}_{0,n} \otimes \mathbf{C}$ can be written as $\mathbf{C}_n = \mathbf{R}_{0,n} \oplus i\mathbf{R}_{0,n}$. Arbitrary $\lambda \in \mathbf{C}_n$ may be written as $\lambda = a + ib$, $a, b \in \mathbf{R}_{0,n}$, leading to the conjugation $\bar{\lambda} = \bar{a} - i\bar{b}$, where the bar denotes the usual Clifford conjugation in $\mathbf{R}_{0,n}$. This leads to the inner product and its associated norm in $\mathbf{C}_n$ given by $(\lambda, \mu) = [\bar{\lambda}\mu]_0$ and $|\lambda| = \sqrt{[\bar{\lambda}\lambda]_0} = (\sum_{\mathscr{A}} |\lambda_{\mathscr{A}}|^2)^{\frac{1}{2}}$.

The vector-valued first order differential operator $\mathscr{D} = \sum_{j=1}^n e_j \partial_{x_j}$ is called the Dirac operator and we have $\mathscr{D}^2 = -\Delta$, where Δ is the Laplace operator in the Euclidean space $\mathbf{R}^n$.

Let Ω be a subdomain of $\mathbf{R}^n$. In what follows, we denote the interior of Ω by Ω_+ and its boundary by $\partial\Omega$. Continuity, Hölder-continuity, continuously differentiability, $\mathscr{L}_p$ $(1 < p < +\infty)$-integral and so on, are defined for a $\mathbf{C}_n$-valued function $\phi = \sum_{\mathscr{A}} \phi_{\mathscr{A}} e_{\mathscr{A}} : \Omega \to \mathbf{C}_n$ where $\phi_{\mathscr{A}} : \Omega \to \mathbf{C}$, by being ascribed to each component $\phi_{\mathscr{A}}$. The corresponding spaces are denoted, respectively, by $\mathscr{C}(\Omega, \mathbf{C}_n)$, $\mathbf{H}^\mu(\Omega, \mathbf{C}_n)$ $(0 < \mu \le 1)$, $\mathscr{C}^1(\Omega, \mathbf{C}_n)$, $\mathscr{L}_p(\Omega, \mathbf{C}_n)$ $(1 < p < +\infty)$ and so on. Null solutions of the Dirac operator $\mathscr{D}$, that is, $\mathscr{D}\phi = 0$, are called (left-) monogenic functions. They are called right-monogenic functions in case where the Dirac operator is applied from the right. The set of left-monogenic functions in Ω forms a right-module, denoted by $\mathbf{M}_{(r)}(\Omega, \mathbf{C}_n)$.

3 Several Lemmas

In this section we present several lemmas which are necessary for the proofs in the next sections.

Definition 1 Suppose Ω is a star-like subdomain of $\mathbf{R}^n$ with the center $a \in \mathbf{R}^n$. The family of Euler operators with a shift on Ω is defined by

$$E_s = sI + \sum_{j=1}^n (x_j - a_j)\partial_{x_j} \quad (s > 0),\ x \in \Omega,$$

where $a = \sum_{j=1}^n e_j a_j$ and I denotes the identity operator on Ω.

The operator $I_s : \mathscr{C}(\Omega, \mathbf{C}_n) \to \mathscr{C}(\Omega, \mathbf{C}_n)$ is defined by

$$I_s\phi = \int_0^1 \phi\big(a + t(x-a)\big)t^{s-1}\,dt \quad (s > 0).$$

Particularly, when Ω is the unit ball with the center at the origin of $\mathbf{R}^n$ we have

$$I_s\phi = \int_0^1 \phi(tx)t^{s-1}\,dt \quad (s > 0),\ x \in \Omega.$$

Moreover, if $\phi \in \mathscr{C}^1(\Omega, \mathbf{C}_n)$ we have $E_s I_s = I_s E_s = I$, $\mathscr{D}E_s\phi = E_{s+1}\mathscr{D}\phi$, $E_s(x-a)\phi = (x-a)E_{s+1}\phi$, where I denotes the identity operator on the space $\mathscr{C}^1(\Omega, \mathbf{C}_n)$. If $\phi \in \mathscr{C}^k(\Omega, \mathbf{C}_n)$ $(k \in \mathbf{N})$ is a solution to the equation $\mathscr{D}^k\phi = 0$, then $E_s\phi$ and $I_s\phi$ are both solutions to the equation $\mathscr{D}^k\phi = 0$, where $\mathscr{D}^k\phi = \mathscr{D}^{k-1}(\mathscr{D}\phi)$. In this context, to avoid discussions of the well-known case, we always assume that $k \geq 2$ in the future.

First of all we need the following decomposition theorem.

Lemma 1 ([30]) *Suppose Ω is a star-like subdomain of $\mathbf{R}^n$ with the center $a \in \mathbf{R}^n$. If the function $\phi \in \mathscr{C}^k(\Omega, \mathbf{C}_n)$ is a solution to the equation $\mathscr{D}^k\phi = 0$, then there exist uniquely determined functions ϕ_j, such that*

$$\phi = \sum_{j=0}^{k-1}(x-a)^j\phi_j, \quad x \in \Omega,$$

where ϕ_j is monogenic on Ω $(j = 0, 1, 2, \ldots, k-1)$.

Secondly, the action of the Dirac operator in the components is given by the following lemma.

Lemma 2 *Let Ω be a subdomain of $\mathbf{R}^n$ with $0 \in \Omega$ and $j \in \mathbf{N}$. If the function $\phi \in \mathscr{C}^1(\Omega, \mathbf{C}_n)$ is monogenic, then*

$$\mathscr{D}x^j\phi = \begin{cases} -2mx^{2m-1}\phi, & \text{if } j = 2m,\\ -2x^{2(m-1)}E_{\frac{n+1}{2}+[\frac{j}{2}]-1}\phi, & \text{if } j = 2m-1, \end{cases}$$

where $x \in \Omega$ and $m \in \mathbf{N}$. Moreover, for $l, p \in \mathbf{N}$ and $2 \leq l \leq j$,

$$\mathscr{D}^l x^j\phi = C_{l,j}x^{j-l}E_{\frac{n+1}{2}+[\frac{j-l}{2}]}\cdots E_{\frac{n+1}{2}+[\frac{j}{2}]-1}\phi, \tag{1}$$

with

$$[s] = \begin{cases} q, & \text{if } q \in \mathbf{N},\\ q+1, & \text{if } s = q+t,\ q \in \mathbf{N},\ 0 < t < 1, \end{cases}$$

and

$$C_{l,j} = \begin{cases} 2^l m(m-1)\dots(m-p+1), & \text{if } j = 2m,\ l = 2p, \\ -2^l m(m-1)\dots(m-p), & \text{if } j = 2m,\ l = 2p+1, \\ 2^l (m-1)\dots(m-p), & \text{if } j = 2m-1,\ l = 2p, \\ -2^l (m-1)\dots(m-p+1), & \text{if } j = 2m-1,\ l = 2p-1. \end{cases}$$

Especially, for $l = j$, we obtain

$$\mathscr{D}^j x^j \phi = E_{\frac{n+1}{2}} \dots E_{\frac{n+1}{2}+m-1}\phi \quad \text{with } C_{l,l} = \begin{cases} 2^j m!, & \text{if } j = 2m, \\ -2^j (m-1)!, & \text{if } j = 2m-1. \end{cases}$$

Proof Applying the well-known relationships

$$\Delta x + x\Delta = -2E_{\frac{n}{2}}, \qquad E\mathscr{D} - \mathscr{D}E = -\mathscr{D}, \qquad Ex - xE = x,$$

with the Euler operator $E = \sum_{j=1}^n x_j \partial_{x_j}$, by direct calculations, the result holds. □

More related results can also be found in [30].

The corresponding lemma for the case of harmonic functions is given as follows:

Lemma 3 *Let Ω be a star-like subdomain of $\mathbf{R}^n$ with $a \in \Omega$ and $j, l \in \mathbf{N}$. If the function $\phi \in \mathscr{C}^2(\Omega, \mathbf{C}_n)$ is harmonic, i.e. null solution to the Laplace operator Δ, then for $l \leq j$,*

$$\Delta^l |x|^{2j}\phi = D_{l,j}|x|^{2(j-l)} E_{\frac{n+1}{2}+j-l} \cdots E_{\frac{n+1}{2}+j-1}\phi, \quad x \in \Omega,$$

where $D_{l,j} = (-4)^l j(j-1)\dots(j-l+1)$ for $0 \leq l \leq j \leq k-1$.

Especially, when $j = l$, for arbitrary $x \in \Omega$, we get

$$\Delta^j |x|^{2j}\phi = (-4)^j j! E_{\frac{n+1}{2}} \dots E_{\frac{n+1}{2}+j-1}\phi.$$

Proof Again, applying the following well-known relationships

$$Ex^2 - x^2E = 2x^2, \qquad \Delta E - E\Delta = -2\Delta, \qquad \Delta x^2 - x^2\Delta = -4E_{\frac{n}{2}},$$

with the Euler operator $E = \sum_{j=1}^n x_j \partial_{x_j}$, by calculating reductively, the result establishes. □

4 Half Dirichlet Problems for Null-Solutions to $\mathscr{D}^k$

In this section, we will present and consider the half Dirichlet problem on the unit ball with boundary values given by $\mathscr{L}_p$-integrable functions in Clifford analysis. Applying an Almansi-type decomposition for iterated Dirac operator in Clifford analysis and associating our shifted Euler operators with the Clifford-Cauchy transform we get the integral representation of the unique solution to it.

Let us denote the open unit ball centered at the origin by $B(1)$ whose closure is $\overline{B}(1)$, and its boundary is S^{n-1}. Furthermore, we denote by B_+ and B_- the interior and exterior of the unit ball. We remark that $\omega \in S^{n-1}$ is the outward pointed unit normal vector of S^{n-1}. Functions taking values in $\mathbf{C}_n$ defined on $B(1)$ and S^{n-1} will be considered, respectively.

We start by introducing the following functions

$$\alpha(x) = \frac{1}{2}(1 + ix), \qquad \beta(x) = \frac{1}{2}(1 - ix), \quad x \in \mathbf{R}^n.$$

Particularly, when $x = \omega \in S^{n-1}$, $\alpha(\omega) = \frac{1}{2}(1 + i\omega)$ and $\beta(\omega) = \frac{1}{2}(1 - i\omega)$.

For arbitrary $f \in \mathscr{L}_p(S^{n-1}, \mathbf{C}_n)$, we define the Cauchy type integral operator

$$\phi(x) = \int_{S^{n-1}} E(\omega - x)\, d\sigma_\omega f(\omega) = (\mathsf{C} f)(x), \quad x \notin S^{n-1}, \tag{2}$$

where $E(\omega - x) = \frac{1}{w_n} \frac{\overline{\omega - x}}{|\omega - x|^n}$ and w_n is the area of the unit sphere of $\mathbf{R}^n$. Then it follows $\mathscr{D}\phi(x) = 0$ and

$$\begin{aligned}\phi^{\pm}(t) = \lim_{x \to t \in S^{n-1}} \phi(x) &= \pm \frac{1}{2} f(t) + \int_{S^{n-1}} E(\omega - t)\, d\sigma_\omega f(\omega)\\ &= \pm \frac{1}{2} f(t) + (\mathscr{H} f)(t),\end{aligned}$$

where $x \in B_\pm$ and the boundary values exit in sense of non-tangential limit. Moreover, we introduce the operator

$$\Phi(x) = \begin{cases} \int_{S^{n-1}} E(\omega - x)\, d\sigma_\omega f(\omega) = (\mathsf{C} f)(x), & x \in B_+,\\ \frac{1}{2} f(t) + \int_{S^{n-1}} E(\omega - t)\, d\sigma_\omega f(\omega) = \frac{1}{2} f(t) + (\mathscr{H} f)(t), & x \in S^{n-1}. \end{cases}$$

Then $\Phi \in \mathscr{L}_p(\overline{B}(1), \mathbf{C}_n)$.

Lemma 4 *If $f \in \mathscr{L}_p(S^{n-1}, \mathbf{C}_n)$ and ϕ as above, then $(I_s\phi)(x)$, $x \notin S^{n-1}$, is well-defined and its boundary value $(I_s\phi)^+(t)$, $t \in S^{n-1}$, exists and $(I_s\phi)^+ \in \mathscr{L}_p(S^{n-1}, \mathbf{C}_n)$.*

Proof For arbitrary $x \notin S^{n-1}$, $s > 0$,

$$(I_s\phi)(x) = \int_0^1 \phi(tx) t^{s-1}\, dt \quad \text{is continuous.}$$

Furthermore, for arbitrary $x \in S^{n-1}$, $s > 0$ and $0 < \varepsilon < 1$, we have

$$(I_s\phi)(x) = \int_0^1 \phi(tx) t^{s-1}\, dt = \int_0^{1-\varepsilon} \phi(tx) t^{s-1}\, dt + \int_{1-\varepsilon}^1 \phi(tx) t^{s-1}\, dt \quad \text{exists.}$$

Hence, for arbitrary $t \in S^{n-1}$, we get that $(I_s\phi)^+(t) = \lim_{x \to t} (I_s\phi)(x)$, $x \in B_+$ exists. Furthermore, for arbitrary $x \in S^{n-1}$, we have

$$\int_{S^{n-1}} (I_s\phi)^+(x)\, dx = \int_0^1 \int_{S^{n-1}} \phi(tx)\, dx\, t^{s-1}\, dt = \int_0^1 \int_{tS^{n-1}} \phi(u)\, du\, t^{s-2}\, dt.$$

Using $\phi(x) \in \mathscr{L}^p(\overline{B}(1), \mathbf{C}_n)$, we get

$$\int_{S^{n-1}} \left|(I_s\phi)^+(x)\right|^p dx \leq \frac{w_n}{s+n-2}\left(\int_{\overline{B}(1)} |\phi(u)|^p du\right)^{\frac{1}{p}}.$$

Therefore, $(I_s\phi)^+ \in \mathscr{L}^p(S^{n-1}, \mathbf{C}_n)$. It follows the result. □

Lemma 5

(i) *If $\phi \in \mathscr{C}^1(B_+, \mathbf{C}_n)$, then $(E_s\phi)(x)$, $x \notin S^{n-1}$, is well-defined.*

(ii) *If $\phi \in \mathscr{C}^k(B_+, \mathbf{C}_n)$ is a solution to $\mathscr{D}^k\phi = 0$ and $\mathscr{D}^l\phi$ $(l = 0, 1, \ldots, k-1)$ has boundary value in the sense of $\mathscr{L}_p$, then for $1 \leq l \leq j \leq k-1$, the boundary value $(E_{\frac{n+1}{2}+[\frac{j-l}{2}]} \cdots E_{\frac{n+1}{2}+[\frac{j}{2}]-1}\phi_j)^+(t)$ of $(E_{\frac{n+1}{2}+[\frac{j-l}{2}]} \cdots E_{\frac{n+1}{2}+[\frac{j}{2}]-1}\phi_j)(x)$ exits on the boundary S^{n-1} from B_+, where $\phi = \sum_{j=0}^{k-1} x^j\phi_j$ with $0 \in B_+$ and ϕ_j are monogenic in B_+ for $j = 0, 1, 2, \ldots, k-1$.*

Proof (i) Since $\phi \in \mathscr{C}^1(B_\pm, \mathbf{C}_n)$, then for arbitrary $x \in B_\pm$ and $s > 0$,

$$E_s\phi(x) = s\phi(x) + \sum_{j=1}^{n} x_j\partial_{x_j}\phi(x) \quad \text{is well-defined and continuous.}$$

(ii) Since $\phi \in \mathscr{C}^k(B_+, \mathbf{C}_n)$ is a solution to $\mathscr{D}^k\phi = 0$ and $0 \in B_+$, by applying Lemma 1, we have

$$\phi = \sum_{j=1}^{k-1} x^j\phi_j, \quad x \in B_+,$$

where ϕ_j is monogenic on B_+ for $j = 0, 1, 2, \ldots, k-1$.

As $\mathscr{D}^l\phi$ $(l = 0, 1, \ldots, k-1)$ has boundary values from B_+ in the sense of $\mathscr{L}_p$, by making use of Lemma 2, we get for $1 \leq l \leq j \leq k-1$,

$$(\mathscr{D}^l\phi)^+(t) = \sum_{j=l}^{k-1} C_{l,j}t^{j-l}(E_{\frac{n+1}{2}+[\frac{j-l}{2}]} \cdots E_{\frac{n+1}{2}+[\frac{j}{2}]-1}\phi_j)^+(t), \quad t \in S^{n-1},$$

where the coefficients $C_{l,j}$ are given by Lemma 2.

Therefore, for $1 \leq l \leq j \leq k-1$, the boundary value $(E_{\frac{n+1}{2}+[\frac{j-l}{2}]} \cdots E_{\frac{n+1}{2}+[\frac{j}{2}]-1}\phi_j)^+(t)$ of $(E_{\frac{n+1}{2}+[\frac{j-l}{2}]} \cdots E_{\frac{n+1}{2}+[\frac{j}{2}]-1}\phi_j)(x)$ exits on the boundary S^{n-1} from B_+ in the sense of $\mathscr{L}_p$. The proof of the result is completed. □

Half-Dirichlet problem (HDP) *Given the boundary data $f_j \in \mathscr{L}^p(S^{n-1}, \mathbf{C}_n)$ $(j = 0, 1, 2, \ldots, k-1)$, find a function $\phi \in \mathscr{C}^k(B_+, \mathbf{C}_n)$ such that*

$$\text{(i)}\quad \begin{cases} \mathscr{D}^k\phi(x)=0, & x\in B_+,\\ \alpha(t)\phi(t)=\alpha(t)f_0(t), & t\in S^{n-1},\\ \alpha(t)(\mathscr{D}\phi)(t)=\alpha(t)f_1(t), & t\in S^{n-1},\\ \quad\vdots & \\ \alpha(t)\big(\mathscr{D}^{k-1}\phi\big)(t)=\alpha(t)f_{k-1}(t), & t\in S^{n-1}, \end{cases}$$

$$\text{(ii)}\quad \begin{cases} \mathscr{D}^k\phi(x)=0, & x\in B_+,\\ \beta(t)\phi(t)=\beta(t)f_0(t), & t\in S^{n-1},\\ \beta(t)(\mathscr{D}\phi)(t)=\beta(t)f_1(t), & t\in S^{n-1},\\ \quad\vdots & \\ \beta(t)\big(\mathscr{D}^{k-1}\phi\big)(t)=\beta(t)f_{k-1}(t), & x\in S^{n-1}, \end{cases}$$

where $\alpha(\omega)=\frac{1}{2}(1+i\omega)$ *and* $\beta(\omega)=\frac{1}{2}(1-i\omega)$.

Theorem 1 *Under the stated conditions, problem* **HDP** (i) *is uniquely solvable and its solution is given by*

$$\phi(x)=\sum_{j=0}^{k-1}x^j\phi_j \tag{3}$$

with

$$\phi_j(x)=\begin{cases} (\mathsf{C}2\alpha\widetilde{f_0})(x), & \text{if } j=0,\\ C_{1,1}^{-1}(I_{\frac{n+1}{2}}\mathsf{C}2\alpha\widetilde{f_1})(x), & \text{if } j=1,\\ C_{l,l}^{-1}(I_{\frac{n+1}{2}+[\frac{l}{2}]-1}\cdots I_{\frac{n+1}{2}}\mathsf{C}2\alpha\widetilde{f_l})(x), & \text{if } 2\le j\le k-1, \end{cases}$$

where $(\mathsf{C}2\alpha\widetilde{f_j})$ $(j=0,1,2,\ldots,k-1)$ *are defined in a similar way as* $(\mathsf{C}f)$ *above and*

$$\widetilde{f_j}(x)=\begin{cases} f_0(x)-\sum_{j=1}^{k-1}C_{j,j}^{-1}x^j(I_{\frac{n+1}{2}+[\frac{j}{2}]-1}\cdots I_{\frac{n+1}{2}}\mathsf{C}2\alpha\widetilde{f_j})(x),\\ \quad \text{if } j=0,\\ f_1(x)-C_{1,2}C_{2,2}^{-1}x(I_{\frac{n+1}{2}}\mathsf{C}2\alpha\widetilde{f_2})(x)-\cdots\\ \qquad -C_{1,k-1}C_{2,2}^{-1}x^{k-2}(I_{\frac{n+1}{2}+[\frac{k-1}{2}]-1}\cdots I_{\frac{n+1}{2}}\mathsf{C}2\alpha\widetilde{f_{k-1}})(x),\\ \quad \text{if } j=1,\ k \text{ odd},\\ f_1(x)-C_{1,2}C_{2,2}^{-1}x(I_{\frac{n+1}{2}}\mathsf{C}2\alpha\widetilde{f_2})(x)-\cdots\\ \qquad -C_{1,k-1}C_{k-1,k-1}^{-1}x^{k-2}(I_{\frac{n+1}{2}+[\frac{k-1}{2}]}\cdots I_{\frac{n+1}{2}}\mathsf{C}2\alpha\widetilde{f_{k-1}})(x),\\ \quad \text{if } j=1,\ k \text{ even},\\ f_l(x)-\sum_{j=l+1}^{k-1}C_{k-1,j}C_{j,j}^{-1}x^j(I_{\frac{n+1}{2}+[\frac{j-l}{2}]}\cdots I_{\frac{n+1}{2}}\mathsf{C}2\alpha\widetilde{f_j})(x),\\ \quad \text{if } 2\le j\le k-1. \end{cases}$$

Proof Since $\mathscr{D}^k\phi(x)=0$ in B_+, by Lemma 1 there exist unique functions ϕ_j satisfying $\mathscr{D}\phi_j=0$ $(j=0,1,2,\ldots,k-1)$ and $\phi=\sum_{j=0}^{k-1}x^j\phi_j$.

Using Lemma 2 for $l\in\mathbf{N}$, $l\le j$, we have

$$\mathscr{D}^l\phi=\sum_{j=0}^{k-1}\mathscr{D}^l\left(x^j\phi_j\right)=\sum_{j=l}^{k-1}C_{l,j}x^{j-l}(E_{\frac{n+1}{2}+[\frac{j-l}{2}]}\cdots E_{\frac{n+1}{2}+[\frac{j}{2}]-1}\phi_j). \tag{4}$$

Then, problem **HDP** (i) is equivalent to the following case

$$(\wp)\quad\begin{cases}\mathscr{D}\phi_j(x)=0\ (j=0,1,2,\ldots,k-1), & x\in B_+,\\ \alpha(x)\sum_{j=0}^{k-1}x^j\phi_j=\alpha(x)f_0(x), & x\in S^{n-1},\\ \alpha(x)\sum_{j=1}^{k-1}C_{1,j}x^{j-1}(E_{\frac{n+1}{2}+[\frac{j-1}{2}]}\cdots E_{\frac{n+1}{2}+[\frac{j}{2}]-1}\phi_j)(x) & \\ \quad=\alpha(x)f_1(x), & x\in S^{n-1},\\ \quad\vdots & \\ \alpha(x)\sum_{j=l}^{k-1}C_{l,j}x^{j-l}(E_{\frac{n+1}{2}+[\frac{j-l}{2}]}\cdots E_{\frac{n+1}{2}+[\frac{j}{2}]-1}\phi_j)(x) & \\ \quad=\alpha(x)f_l(x), & x\in S^{n-1},\\ \quad\vdots & \\ \alpha(x)C_{k-1,k-1}(E_{\frac{n+1}{2}}\ldots E_{\frac{n+1}{2}+[\frac{k-1}{2}]-1}\phi_{k-1})(x) & \\ \quad=\alpha(x)f_{k-1}(x), & x\in S^{n-1}.\end{cases}$$

We now proceed inductively. First, we consider the system

$$(*)\quad\begin{cases}\mathscr{D}\phi_{k-1}(x)=0, & x\in B_+,\\ \alpha(x)C_{k-1,k-1}(E_{\frac{n+1}{2}}\ldots E_{\frac{n+1}{2}+[\frac{k-1}{2}]-1}\phi_{k-1})(x) & \\ \quad=\alpha(x)f_{k-1}(x), & x\in S^{n-1}.\end{cases}$$

By Lemma 1 problem $(*)$ has the unique solution

$$\begin{aligned}\phi_{k-1}(x)&=C_{k-1,k-1}^{-1}I_{\frac{n+1}{2}+[\frac{k-1}{2}]-1}\cdots I_{\frac{n+1}{2}}\int_{S^{n-1}}E(\omega-x)\,d\sigma_\omega 2\alpha(\omega)f_{k-1}(\omega)\\ &=C_{k-1,k-1}^{-1}(I_{\frac{n+1}{2}+[\frac{k-1}{2}]-1}\cdots I_{\frac{n+1}{2}}\mathrm{C}2\alpha\widetilde{f}_{k-1})(x),\quad x\in B_+,\end{aligned}\tag{5}$$

with $\widetilde{f}_{k-1}\equiv f_{k-1}(\omega)$ in S^{n-1}.

Using Lemma 4 we get that

$$\begin{aligned}&(I_{\frac{n+1}{2}+[\frac{k-1}{2}]-1}\cdots I_{\frac{n+1}{2}}\mathrm{C}2\alpha\widetilde{f}_{k-1})(x)\\ &=\lim_{x\to t}I_{\frac{n+1}{2}+[\frac{k-1}{2}]-1}\cdots I_{\frac{n+1}{2}}\int_{S^{n-1}}E(\omega-x)\,d\sigma_\omega\mathrm{C}2\alpha f_{k-1}(\omega),\quad x\in B_+,\end{aligned}$$

belongs to $\mathscr{L}_p(S^{n-1},\mathbf{C}_{n+1})$. Therefore, $\phi_{k-1}\in\mathscr{L}_p(S^{n-1},\mathbf{C}_n)$ thus, implying also $t\phi_{k-1}\in\mathscr{L}_p(S^{n-1},\mathbf{C}_n)$.

Then, we consider the second boundary value problem

$$(**)\quad \begin{cases} \mathscr{D}\phi_{k-2}(x) = 0, & x \in B_+, \\ \alpha(x) C_{k-2,k-2}(E_{\frac{n+1}{2}} \ldots E_{\frac{n+1}{2}+[\frac{k-2}{2}]-1}\phi_{k-2})(x) & \\ \quad = \alpha(x)\widetilde{f}_{k-2}(x), & x \in S^{n-1}, \end{cases}$$

where we identify $\widetilde{f}_{k-2} \equiv f_{k-2} - C_{k-1,k-2}C^{-1}_{k-1,k-1}x(I_{\frac{n+1}{2}}\mathrm{C}2\alpha f_{k-1})$ in S^{n-1}. Hence, problem $(**)$ has the unique solution

$$\begin{aligned} \phi_{k-2}(x) &= C^{-1}_{k-2,k-2} I_{\frac{n+1}{2}+[\frac{k-2}{2}]-1} \cdots I_{\frac{n+1}{2}} \int_{S^{n-1}} E(\omega - x)\, d\sigma_\omega 2\alpha(\omega)\widetilde{f}_{k-2}(\omega) \\ &= C^{-1}_{k-2,k-2}(I_{\frac{n+1}{2}+[\frac{k-2}{2}]-1} \cdots I_{\frac{n+1}{2}}\mathrm{C}2\alpha\widetilde{f}_{k-2})(x), \quad x \in B_+. \end{aligned} \tag{6}$$

By induction on $2 \le l \le k-1$, the following boundary value problem

$$\begin{cases} \mathscr{D}\phi_l(x) = 0, & x \in B_+, \\ \alpha(x)C_{l,l}(E_{\frac{n+1}{2}} \ldots E_{\frac{n+1}{2}+[\frac{l}{2}]-1}\phi)(x) = \alpha(x)\widetilde{f}_l(x), & x \in S^{n-1}, \end{cases}$$

where we identify $\widetilde{f}_l \equiv f_l - \sum_{j=l+1}^{k-1} C_{k-1,j}C^{-1}_{j,j}x^j(I_{\frac{n+1}{2}+[\frac{j-l}{2}]} \cdots I_{\frac{n+1}{2}}\mathrm{C}2\alpha\widetilde{f}_j)$ in S^{n-1}, has the unique solution

$$\begin{aligned} \phi_l(x) &= C^{-1}_{l,l} I_{\frac{n+1}{2}+[\frac{l}{2}]-1} \cdots I_{\frac{n+1}{2}} \int_{S^{n-1}} E(\omega - x)\, d\sigma_\omega 2\alpha(\omega)\widetilde{f}_l(\omega) \\ &= C^{-1}_{l,l}(I_{\frac{n+1}{2}+[\frac{l}{2}]-1} \cdots I_{\frac{n+1}{2}}\mathrm{C}2\alpha\widetilde{f}_l)(x), \quad x \in B_+. \end{aligned} \tag{7}$$

The remaining two cases ($l = 0, 1$) are treated as follows. For $l = 1$, the boundary value problem

$$\begin{cases} \mathscr{D}\phi_1(x) = 0, & x \in B_+, \\ \alpha(x)C_{1,1}(E_{\frac{n+1}{2}}\phi_1)(x) = \alpha(x)\widetilde{f}_1(x), & x \in S^{n-1}, \end{cases}$$

where $\widetilde{f}_1$ is identified, in S^{n-1}, with

$$\begin{cases} f_1(t) - C_{1,2}C^{-1}_{2,2}t(I_{\frac{n+1}{2}}\mathrm{C}2\alpha\widetilde{f}_2)(t) - \cdots & \\ \quad - C_{1,k-1}C^{-1}_{k-1,k-1}t^{k-2}(I_{\frac{n+1}{2}+[\frac{k-1}{2}]-1} \cdots I_{\frac{n+1}{2}}\mathrm{C}2\alpha\widetilde{f}_{k-1})(t), & \text{if } k \text{ odd}, \\ & \\ f_1(t) - C_{1,2}C^{-1}_{2,2}t(I_{\frac{n+1}{2}}\mathrm{C}2\alpha\widetilde{f}_2)(t) - \cdots & \\ \quad - C_{1,k-1}C^{-1}_{k-1,k-1}t^{k-2}(I_{\frac{n+1}{2}+[\frac{k-1}{2}]} \cdots I_{\frac{n+1}{2}}\mathrm{C}2\alpha\widetilde{f}_{k-1})(t), & \text{if } k \text{ even}, \end{cases}$$

has the unique solution

$$\begin{aligned} \phi_1(x) &= C^{-1}_{1,1} I_{\frac{n+1}{2}} \int_{S^{n-1}} E(\omega - x)\, d\sigma_\omega 2\alpha(\omega)\widetilde{f}_1(\omega) \\ &= C^{-1}_{1,1}(I_{\frac{n+1}{2}}\mathrm{C}2\alpha\widetilde{f}_1)(x), \quad x \in B_+. \end{aligned} \tag{8}$$

For $l = 0$, the boundary value problem

$$\begin{cases} \mathscr{D}\phi_1(x) = 0, & x \in B_+, \\ \alpha(x)\phi_0(x) = \alpha(x)\widetilde{f_0}(x), & x \in S^{n-1}, \end{cases}$$

where $\widetilde{f_0} \equiv f_0 - \sum_{j=1}^{k-1} C_{j,j}^{-1} x^j (I_{\frac{n+1}{2}+[\frac{j}{2}]-1} \cdots I_{\frac{n+1}{2}} \mathsf{C}2\alpha \widetilde{f_j})$ in S^{n-1}, has the unique solution

$$\phi_0(x) = \int_{S^{n-1}} E(\omega - x)\, d\sigma_\omega 2\alpha(\omega) \widetilde{f_0}(\omega) = (\mathsf{C}2\alpha \widetilde{f_0})(x), \quad x \in B_+. \tag{9}$$

Associating the terms (5)–(9) we have that problem ($\wp$) has the unique solution $\phi(x) = \sum_{j=0}^{k-1} x^j \phi_j$. It follows the result. □

Similarly, we have the following theorem for the second kind of problem. Since its proof follows the same lines as the previous theorem, it will be omitted here.

Theorem 2 *Under the stated conditions, problem* **HDP** (ii) *is uniquely solvable and its solution is given by*

$$\phi(x) = \sum_{j=0}^{k-1} x^j \phi_j(x) \tag{10}$$

where

$$\phi_j(x) = \begin{cases} (\mathsf{C}2\beta \widetilde{f_0})(x), & \text{if } j = 0, \\ C_{1,1}^{-1}(I_{\frac{n+1}{2}} \mathsf{C}2\beta \widetilde{f_1})(x), & \text{if } j = 1, \\ C_{l,l}^{-1}(I_{\frac{n+1}{2}+[\frac{l}{2}]-1} \cdots I_{\frac{n+1}{2}} \mathsf{C}2\beta \widetilde{f_l})(x), & \text{if } 2 \le j \le k-1, \end{cases}$$

the transforms $(\mathsf{C}2\beta \widetilde{f_j})$ $(j = 0, 1, 2, \ldots, k-1)$ *are defined similarly to* $(\mathsf{C}f)$ *and*

$$\widetilde{f_j}(x) = \begin{cases} f_0(x) - \sum_{j=1}^{k-1} C_{j,j}^{-1} x^j (I_{\frac{n+1}{2}+[\frac{j}{2}]-1} \cdots I_{\frac{n+1}{2}} \mathsf{C}2\beta \widetilde{f_j})(x), \\ \quad \text{if } j = 0, \\ f_1(x) - C_{1,2} C_{2,2}^{-1} x (I_{\frac{n+1}{2}} \mathsf{C}2\beta \widetilde{f_2})(x) - \cdots \\ \qquad - C_{1,k-1} C_{2,2}^{-1} x^{k-2} (I_{\frac{n+1}{2}+[\frac{k-1}{2}]-1} \cdots I_{\frac{n+1}{2}} \mathsf{C}2\beta \widetilde{f}_{k-1})(x), \\ \quad \text{if } j = 1,\ k \text{ odd}, \\ f_1(x) - C_{1,2} C_{2,2}^{-1} x (I_{\frac{n+1}{2}} \mathsf{C}2\beta \widetilde{f_2})(x) - \cdots \\ \qquad - C_{1,k-1} C_{k-1,k-1}^{-1} x^{k-2} (I_{\frac{n+1}{2}+[\frac{k-1}{2}]} \cdots I_{\frac{n+1}{2}} \mathsf{C}2\beta \widetilde{f}_{k-1})(x), \\ \quad \text{if } j = 1,\ k \text{ even}, \\ f_l(x) - \sum_{j=l+1}^{k-1} C_{k-1,j} C_{j,j}^{-1} x^j (I_{\frac{n+1}{2}+[\frac{j-l}{2}]} \cdots I_{\frac{n+1}{2}} \mathsf{C}2\beta \widetilde{f_j})(x), \\ \quad \text{if } 2 \le j \le k-1. \end{cases}$$

Remark 1 Other kind of half Dirichlet problems can also be discussed in a similar way as in **HDP** (i) and (ii). When $k = 2l$, $l \in \mathbf{N}$, problems **HDP** (i) and (ii) reduce to the following half Dirichlet problems for poly-harmonic functions, respectively,

$$\textbf{HDP (iii)} \quad \begin{cases} \Delta^l \phi(x) = 0, & x \in B_+, \\ \alpha(t)\phi(t) = \alpha(t) f_0(t), & t \in S^{n-1}, \\ \alpha(t)(\mathscr{D}\phi)(t) = \alpha(t) f_1(t), & t \in S^{n-1}, \\ \quad \vdots & \\ \alpha(t)\big(\mathscr{D}^{2l-1}\phi\big)(t) = \alpha(t) f_{k-1}(t), & t \in S^{n-1}, \end{cases}$$

$$\textbf{HDP (iv)} \quad \begin{cases} \Delta^l \phi(x) = 0, & x \in B_+, \\ \beta(t)\phi(t) = \beta(t) f_0(t), & t \in S^{n-1}, \\ \beta(t)(\mathscr{D}\phi)(t) = \beta(t) f_1(t), & t \in S^{n-1}, \\ \quad \vdots & \\ \beta(t)\big(\mathscr{D}^{2l-1}\phi\big)(t) = \beta(t) f_{k-1}(t), & x \in S^{n-1}. \end{cases}$$

Remark 2 For $k = 1$, problems **HDP** (i) and (ii) have been treated already in [10, 20]. This implies that problems **HDP** (iii) and (iv) on the unit ball can be solved by means of Clifford analysis.

5 Dirichlet Problems for Poly-harmonic Functions

In this section we consider the Dirichlet problem for poly-harmonic functions on the unit ball with boundary data given by $\mathscr{L}_p$-integral functions in Clifford analysis. We use the Almansi decomposition theorem for poly-harmonic functions and Clifford-integral operators in order to express its unique solution in two different ways.

The Poisson kernel (see e.g. [31]) on the unit ball of $\mathbf{R}^n$ is defined as

$$P(x,\omega) = \frac{1-|x|^2}{|x-\omega|^n}, \quad x \in B_+,\ \omega \in S^{n-1}.$$

For an arbitrary $f \in \mathscr{L}_p(S^{n-1}, \mathbf{C}_n)$ we introduce its Poisson integral transform as

$$(\mathbf{P}f)(x) = \int_{S^{n-1}} P(x,\omega) f(\omega)\, d\sigma_\omega, \quad x \in B_+. \tag{11}$$

We extend $\mathbf{P}f$ to the unit sphere by setting $(\mathbf{P}f)(t) = \lim_{B_+ \ni x \to t \in S^{n-1}} (\mathbf{P}f)(x)$. Moreover, we have that $\lim_{B_+ \ni x \to t \in S^{n-1}} (\mathbf{P}f)(x) = f(t)$, $\mathbf{P}f$ is harmonic on B_+ and $\mathbf{P}f \in \mathscr{L}_p(\overline{B}(1), \mathbf{C}_n)$.

Lemma 6 *Let $f \in \mathscr{L}_p(S^{n-1}, \mathbf{C}_n)$. Then $(I_s\mathbf{P}f)(x)$, $x \in B_+$ is well-defined, its boundary value $(I_s\mathbf{P}f)^+(t) = \lim_{x \to t \in S^{n-1}} (I_s\mathbf{P}f)(x)$ exists for all $t \in S^{n-1}$, and $(I_s\mathbf{P}f)^+ \in \mathscr{L}_p(S^{n-1}, \mathbf{C}_n)$.*

Proof For arbitrary $x \in B_+$, and $s > 0$ we have that

$$(I_s\mathbf{P}f)(x) = \int_0^1 \mathbf{P}f(tx) t^{s-1}\, dt$$

is continuous. Moreover, the following decomposition holds for $0 < \varepsilon < 1$

$$
\begin{aligned}
(I_s\mathbf{P}f)(x) &= \int_0^1 \mathbf{P}f(tx)t^{s-1}\,dt \\
&= \int_0^{1-\varepsilon} \mathbf{P}f(tx)t^{s-1}\,dt + \int_{1-\varepsilon}^1 \mathbf{P}f(tx)t^{s-1}\,dt \quad \text{exists.}
\end{aligned}
$$

Hence for arbitrary $t \in S^{n-1}$, we get $(I_s\mathbf{P}f)^+(t) = \lim_{x\to t\in S^{n-1}}(I_s\mathbf{P}f)(x),\ x \in B_+$ exists.

Furthermore, we have

$$
\int_{S^{n-1}} (I_s\mathbf{P}f)^+(x)\,dx = \int_0^1 \int_{S^{n-1}} \mathbf{P}f(tx)\,dx\,t^{s-1}\,dt = \int_0^1 \int_{tS^{n-1}} \mathbf{P}f(u)\,du\,t^{s-2}\,dt.
$$

Associating $\mathbf{P}f \in \mathscr{L}_p(\overline{B}(1), \mathbf{C}_n)$, we get

$$
\int_{S^{n-1}} \left|(I_s\mathbf{P}f)^+(x)\right|^p dx \le \frac{w_n}{s+n-2}\left(\int_{\overline{B}(1)} \left|\mathbf{P}f(u)\right|^p du\right)^{\frac{1}{p}}.
$$

Therefore, $(I_s\mathbf{P}f)^+ \in \mathscr{L}_p(S^{n-1}, \mathbf{C}_n)$. It follows the result. □

Lemma 7 *If $\phi \in \mathscr{C}^{2k}(B_+, \mathbf{C}_n)$ is a solution to $\Delta^k\phi = 0$ and $\Delta^l\phi$ $(l = 0, 1, \ldots, k-1)$ has boundary values from B_+ in the sense of $\mathscr{L}_p$, then for $1 \le l \le j \le k-1$, the boundary value $(E_{\frac{n+1}{2}+[\frac{j-l}{2}]} \cdots E_{\frac{n+1}{2}+[\frac{j}{2}]-1}\phi_j)^+(t)$ of $(E_{\frac{n+1}{2}+[\frac{j-l}{2}]} \cdots E_{\frac{n+1}{2}+[\frac{j}{2}]-1}\phi_j)(x)$ exists on the boundary S^{n-1} from B_+, where $\phi = \sum_{j=0}^{k-1} |x|^{2j}\phi_j$ with $0 \in B_+$ and ϕ_j are harmonic in B_+ for $j = 0, 1, 2, \ldots, k-1$.*

Proof Since $\phi \in \mathscr{C}^{2k}(B_+, \mathbf{C}_n)$ is a solution to $\Delta^k\phi = 0$ and $0 \in B_+$, then by Lemma 1, we have

$$
\phi(x) = \sum_{j=0}^{k-1} |x|^{2j}\phi_j(x), \quad x \in B_+,
$$

where ϕ_j is harmonic on B_+ for $j = 0, 1, 2, \ldots, k-1$.

As $\Delta^l\phi$ $(l = 0, 1, \ldots, k-1)$ has boundary values from B_+ in the sense of $\mathscr{L}_p$, by making use of Lemma 3, we get for $1 \le l \le j \le k-1$,

$$
\left(\Delta^l\phi\right)^+(t) = \sum_{j=l}^{k-1} D_{l,j}|t|^{2(j-l)}(E_{\frac{n+1}{2}+[\frac{j-l}{2}]} \cdots E_{\frac{n+1}{2}+[\frac{j}{2}]-1}\phi_j)^+(t), \quad t \in S^{n-1},
$$

where the coefficients $D_{l,j}$ are given by Lemma 3.

Therefore, for $1 \le l \le j \le k-1$, the boundary value $(E_{\frac{n+1}{2}+[\frac{j-l}{2}]} \cdots E_{\frac{n+1}{2}+[\frac{j}{2}]-1}\phi_j)^+(t)$ of $(E_{\frac{n+1}{2}+[\frac{j-l}{2}]} \cdots E_{\frac{n+1}{2}+[\frac{j}{2}]-1}\phi_j)(x)$ exists on the boundary S^{n-1} from B_+ in the sense of $\mathscr{L}_p$. The proof of the result is completed. □

Dirichlet problem (DP) *Given the boundary data* $f_j \in \mathscr{L}_p(S^{n-1}, \mathbf{C}_n)$ $(j = 0, 1, 2, 3, \ldots, k-1)$, *find a function* $\phi \in \mathscr{C}^{2k}(B_+, \mathbf{C}_n)$ *such that*

$$(\star) \quad \begin{cases} \Delta^k \phi(x) = 0, & x \in B_+, \\ \phi(t) = f_0(t), & t \in S^{n-1}, \\ (\Delta\phi)(t) = f_1(t), & t \in S^{n-1}, \\ \vdots & \\ (\Delta^{k-1}\phi)(t) = f_{k-1}(t), & t \in S^{n-1}. \end{cases}$$

Theorem 3 *The problem* $(\star)$ *is solvable and its solution is given via*

$$\phi = \sum_{j=0}^{k-1} |x|^{2j} \phi_j,$$

where for arbitrary $x \in B_+$, ϕ_j *is given by*

$$\phi_l(x) = \begin{cases} D_{k-1,k-1}^{-1}(I_{\frac{n+1}{2}+k-2} \cdots I_{\frac{n+1}{2}} \mathbf{P} \widetilde{f}_l)(x), & \textit{if } l = k-1, \\ D_{l,l}^{-1}(I_{\frac{n+1}{2}+l-1} \cdots I_{\frac{n+1}{2}} \mathbf{P} \widetilde{f}_l)(x), & \textit{if } 0 \le l \le k-2, \end{cases}$$

and, for arbitrary $\omega \in S^{n-1}$, $\widetilde{f}_l$ *is given by*

$$\widetilde{f}_l(\omega) = \begin{cases} f_{k-1}(\omega), \quad \textit{if } l = k-1, \\ f_l(\omega) - \sum_{j=l+1}^{k-1} D_{l+1,j} D_{l,l}^{-1}(I_{\frac{n+1}{2}+j-l-1} \cdots I_{\frac{n+1}{2}} \phi_j)(\omega), \\ \quad \textit{if } 0 \le l \le k-2, \end{cases}$$

and $D_{l,j} = (-4)^l j(j-1)\ldots(j-l+1)$ *for* $0 \le l \le j \le k-1$.

Proof Since $\Delta^l \phi(x) = 0$, $x \in B_+$, by applying Almansi decomposition theorem in [32], there exist unique harmonic functions ϕ_j $(j = 0, 1, 2, \ldots, k-1)$ satisfying

$$\phi = \sum_{j=0}^{k-1} |x|^{2j} \phi_j.$$

Using Lemma 3, for $l \le j \le k-1$, we have

$$\Delta^l |x|^{2j} \phi = D_{l,j} |x|^{2(j-l)} E_{\frac{n+1}{2}+j-l} \cdots E_{\frac{n+1}{2}+j-1} \phi,$$

where $D_{l,j} = (-4)^l j(j-1)\ldots(j-l+1)$. Therefore, problem $(\star)$ is equivalent to

$$\begin{cases} \Delta\phi_j(x)=0, \quad j=0,1,2,\ldots,k-1, & x\in B_+,\\ \sum_{j=0}^{k-1}\phi_j(t)=f_0(t), & t\in S^{n-1},\\ \sum_{j=1}^{k-1}D_{1,j}(E_{\frac{n+1}{2}+j-1}\phi_j)(t)=f_1(t), & t\in S^{n-1},\\ \vdots \\ \sum_{j=l}^{k-1}D_{l,j}(E_{\frac{n+1}{2}+j-l}\cdots E_{\frac{n+1}{2}+j-1}\phi_j)(t)=f_l(t), & t\in S^{n-1},\\ \vdots \\ D_{k-1,k-1}(E_{\frac{n+1}{2}}\ldots E_{\frac{n+1}{2}+k-2}\phi_{k-1})(t)=f_{k-1}(t), & t\in S^{n-1}. \end{cases}$$

Let us consider the following boundary value problem

$$(\aleph)\quad \begin{cases} \Delta\phi_{k-1}(x)=0, & x\in B_+,\\ D_{k-1,k-1}(E_{\frac{n+1}{2}}\ldots E_{\frac{n+1}{2}+k-2}\phi_{k-1})(t)=f_{k-1}(t), & t\in S^{n-1}. \end{cases}$$

By using the operator I_s ($s>0$), problem ($\aleph$) has the unique solution

$$\begin{aligned}\phi_{k-1}(x)&=D_{k-1,k-1}^{-1}I_{\frac{n+1}{2}+k-2}\cdots I_{\frac{n+1}{2}}\int_{S^{n-1}}P(x,\omega)f_{k-1}(\omega)\,d\sigma_\omega\\ &=D_{k-1,k-1}^{-1}(I_{\frac{n+1}{2}+k-2}\cdots I_{\frac{n+1}{2}}\mathbf{P}\widetilde{f}_{k-1})(x),\quad x\in B_+,\end{aligned}\tag{12}$$

where $\widetilde{f}_{k-1}(\omega)=f_{k-1}(\omega)$, $\omega\in S^{n-1}$.

Applying Lemma 6, for arbitrary $x\in B_+$, we get that

$$(I_{\frac{n+1}{2}+k-2}\cdots I_{\frac{n+1}{2}}\mathbf{P}\widetilde{f}_{k-1})(t)=\lim_{x\to t\in S^{n-1}}(I_{\frac{n+1}{2}+k-2}\cdots I_{\frac{n+1}{2}}\mathbf{P}\widetilde{f}_{k-1})(x)$$

belongs to $\mathscr{L}_p(S^{n-1},\mathbf{C}_n)$. Next, we consider the following boundary value problem

$$(\aleph\aleph)\quad \begin{cases} \Delta\phi_{k-l}(x)=0, & x\in B_+,\\ \sum_{j=k-2}^{k-1}D_{l,j}(E_{\frac{n+1}{2}+j-l}\cdots E_{\frac{n+1}{2}+j-1}\phi_j)(t)=f_{k-2}(t), & t\in S^{n-1}. \end{cases}$$

Again, problem ($\aleph\aleph$) has the unique solution

$$\begin{aligned}\phi_{k-2}(x)&=D_{k-2,k-2}^{-1}I_{\frac{n+1}{2}+k-3}\cdots I_{\frac{n+1}{2}}\int_{S^{n-1}}P(x,\omega)\widetilde{f}_{k-2}(\omega)\,d\sigma_\omega\\ &=D_{k-2,k-2}^{-1}(I_{\frac{n+1}{2}+k-3}\cdots I_{\frac{n+1}{2}}\mathbf{P}\widetilde{f}_{k-1})(x),\quad x\in B_+,\end{aligned}\tag{13}$$

where $\widetilde{f}_{k-2}(\omega)=f_{k-1}(\omega)-D_{k-2,k-1}D_{k-2,k-2}^{-1}(I_{\frac{n+1}{2}}\phi_{k-1})(\omega)$, $\omega\in S^{n-1}$.

Now, by recursion, for $0\le l\le k-2$, the boundary value problem

$$\begin{cases} \Delta\phi_{k-l}(x)=0, & x\in B_+,\\ \sum_{j=l}^{k-1}D_{l,j}(E_{\frac{n+1}{2}+j-l}\cdots E_{\frac{n+1}{2}+j-1}\phi_j)(t)=f_l(t), & t\in S^{n-1}, \end{cases}$$

has the unique solution

$$\phi_l(x) = D_{l,l}^{-1} I_{\frac{n+1}{2}+l-1} \cdots I_{\frac{n+1}{2}} \int_{S^{n-1}} P(x,\omega) \widetilde{f}_l(\omega)\, d\sigma_\omega$$
$$= D_{l,l}^{-1} (I_{\frac{n+1}{2}+l-1} \cdots I_{\frac{n+1}{2}} \mathbf{P} \widetilde{f}_l)(x), \quad x \in B_+, \tag{14}$$

where $\widetilde{f}_l(\omega) = f_l(\omega) - \sum_{j=l+1}^{k-1} D_{l+1,j} D_{l,l}^{-1} (I_{\frac{n+1}{2}+j-l-1} \cdots I_{\frac{n+1}{2}} \phi_j)(\omega)$, $\omega \in S^{n-1}$.

From this it follows the result. □

Applying Lemma 4 and references [10, 20, 22] directly, we obtain the following theorem.

Theorem 4 *Problem* ($\star$) *is solvable and its solution can be also expressed by*

$$\phi = \sum_{j=0}^{k-1} |x|^{2j} \phi_j,$$

where for arbitrary $x \in B_+$,

$$\phi_l(x) = \begin{cases} D_{k-1,k-1}^{-1} (I_{\frac{n+1}{2}+k-2} \cdots I_{\frac{n+1}{2}} (\alpha \mathsf{C}2\alpha \widetilde{f}_l + \beta \mathsf{C}2\beta \widetilde{f}_l))(x), & \text{if } l = k-1, \\ D_{l,l}^{-1} (I_{\frac{n+1}{2}+l-1} \cdots I_{\frac{n+1}{2}} (\alpha \mathsf{C}2\alpha \widetilde{f}_l + \beta \mathsf{C}2\beta \widetilde{f}_l))(x), & \text{if } 0 \le l \le k-2, \end{cases}$$

$\mathsf{C}2\alpha \widetilde{f}_l, \mathsf{C}2\beta \widetilde{f}_l$ $(0 \le l \le k-1)$ *are defined similarly to* $\mathsf{C}f$ *and, for arbitrary* $\omega \in S^{n-1}$,

$$\widetilde{f}_l(\omega) = \begin{cases} f_{k-1}(\omega), & \text{if } l = k-1, \\ f_l(\omega) - \sum_{j=l+1}^{k-1} D_{l+1,j} D_{l,l}^{-1} (I_{\frac{n+1}{2}+j-l-1} \cdots I_{\frac{n+1}{2}} \phi_j)(\omega), & \\ \quad \text{if } 0 \le l \le k-2, & \end{cases}$$

and $D_{l,j} = (-4)^l j(j-1)\ldots(j-l+1)$ *for* $0 \le l \le j \le k-1$.

Remark 3 Theorem 3 means that the solution to the Dirichlet problem on the unit ball for iterated Dirac operator of even order is given in view of the Dirichlet problem for harmonic functions, which is a different representation than that of Theorems 1 and 2. Theorem 4 implies that the Dirichlet problem on the unit ball for iterated Dirac operator of even order could be given also solemnly by using the Cauchy transform in Clifford analysis.

Acknowledgements This work was supported by *FEDER* funds through *COMPETE*—Operational Programme Factors of Competitiveness ("Programa Operacional Factores de Competitividade"), by Portuguese funds through the *Center for Research and Development in Mathematics and Applications* (University of Aveiro) and the Portuguese Foundation for Science and Technology ("FCT—Fundação para a Ciência e a Tecnologia"), within project PEst-C/MAT/UI4106/2011 with COMPETE number FCOMP-01-0124-FEDER-022690 and by NNSF of China under Grant No. 60873249. The first author is the recipient of Postdoctoral Foundation from FCT (Portugal) under Grant No. SFRH/BPD/74581/2010 and from China under Grant No. 201003111.

References

1. Balk, M.B.: On Polyanalytic Functions. Akademie Verlag, Berlin (1991)
2. Begehr, H., Du, J., Wang, Y.: A Dirichlet problem for polyharmonic functions. Ann. Mat. Pura Appl. **187**(4), 435–457 (2008)
3. Begehr, H., Du, Z., Wang, N.: Dirichlet problems for inhomogeneous complex mixed-partial differential equations of higher order in the unit disc: new view. Oper. Theory Adv. Appl. **205**, 101–128 (2009)
4. Du, Z., Guo, G., Wang, N.: Decompositions of functions and Dirichlet problems in the unit disc. J. Math. Anal. Appl. **362**(1), 1–16 (2010)
5. Begehr, H., Hile, G.N.: A hierarchy of integral operators. Rocky Mt. J. Math. **27**, 669–706 (1997)
6. Brackx, F., Delanghe, R., Sommen, F.: Clifford Analysis. Research Notes in Mathematics, vol. 76. Pitman, London (1982)
7. Delanghe, R., Sommen, F., Souček, V.: Clifford Algebra and Spinor Valued Functions. Kluwer Academic, Dordrecht (1992)
8. Gilbert, J.E., Murray, M.A.M.: Clifford Algebras and Dirac Operators in Harmonic Analysis. Cambridge Studies in Advances Mathematics, vol. 26. Cambridge University Press, Cambridge (1991)
9. Gürlebeck, K., Sprößig, W.: Quaternionic and Clifford Calculus for Physicists and Engineers. Wiley, Chichester (1997)
10. Delanghe, R., Sommen, F., Xu, Z.: Half Dirichlet problems for powers of the Dirac operator in the unit ball of $\mathbf{R}^m$ $(m \geq 3)$. Bull. Belg. Math. Soc. Simon Stevin **42**(3), 409–429 (1990)
11. Mitrea, M.: Half-Dirichlet problems for Dirac operators in Lipschitz domains. Adv. Appl. Clifford Algebras **11**(1), 121–135 (2001)
12. Delanghe, R., Qian, T.: Half Dirichlet problems and decompositions of Poisson kernels. Adv. Appl. Clifford Algebras **17**(3), 383–393 (2007)
13. Gürlebeck, K., Kähler, U.: On a boundary value problem of the biharmonic equation. Math. Methods Appl. Sci. **20**(4), 867–883 (1997)
14. Cerejeiras, P., Kähler, U.: Elliptic boundary value problems of fluid dynamics over unbounded domains. Math. Methods Appl. Sci. **23**, 81–101 (2000)
15. Bernstein, S.: On the left linear Riemann problem in Clifford analysis. Bull. Belg. Math. Soc. Simon Stevin **3**, 557–576 (1996)
16. Abreu Blaya, R., Bory Reyes, J.: On the Riemann Hilbert type problems in Clifford analysis. Adv. Appl. Clifford Algebras **11**(1), 15–26 (2001)
17. Gong, Y.F., Du, J.Y.: A kind of Riemann and Hilbert boundary value problem for left monogenic functions in $\mathbf{R}^m$ $(m \geq 2)$. Complex Var. **49**(5), 303–318 (2004)
18. Gürlebeck, K., Zhang, Z.Z.: Some Riemann boundary value problems in Clifford analysis. Math. Methods Appl. Sci. **33**, 287–302 (2010)
19. Ku, M., Kähler, U., Wang, D.S.: Riemann boundary value problems on the sphere in Clifford analysis. Adv. Appl. Clifford Algebras (2012). doi:10.1007/s00006-011-0308-2
20. Ku, M., Wang, D.S.: Half Dirichlet problem for matrix functions on the unit ball in Hermitian Clifford analysis. J. Math. Anal. Appl. **374**, 442–457 (2011)
21. Ku, M., Kähler, U.: Riemann boundary value problems on half space in Clifford analysis. Math. Methods Appl. Sci. (2012). doi:10.1002/mma.2557
22. Ku, M., Kähler, U., Wang, D.S.: Half Dirichlet problem for Hölder continuous matrix functions in Hermitian Clifford analysis. Complex Var. Elliptic Equ. (2012). doi: 10.1080/17476933.2011.649738
23. Ku, M.: Integral formula of isotonic functions over unbounded domain in Clifford analysis. Adv. Appl. Clifford Algebras **20**(1), 57–70 (2010)
24. Ryan, J.: Complexified Clifford analysis. Complex Var. Theory Appl. **1**(1), 119–149 (1982)
25. Ku, M., Du, J.Y., Wang, D.S.: On generalization of Martinelli-Bochner integral formula using Clifford analysis. Adv. Appl. Clifford Algebras **20**(2), 351–366 (2010)

26. Ku, M., Du, J.Y., Wang, D.S.: Some properties of holomorphic Cliffordian functions in complex Clifford analysis. Acta Math. Sci. **30B**(3), 747–768 (2010)
27. Ku, M., Wang, D.S.: Solutions to the polynomial Dirac equations on unbounded domains in Clifford analysis. Math. Methods Appl. Sci. **34**(4), 418–427 (2011)
28. Ku, M., Wang, D.S., Dong, L.: Solutions to polynomials generalized Bers-Vekua equations in Clifford analysis. Complex Anal. Oper. Theory **6**(2), 407–424 (2012)
29. Ku, M., Du, J.Y.: On integral representation of spherical k-regular functions in Clifford analysis. Adv. Appl. Clifford Algebras **19**(1), 83–100 (2009)
30. Malonek, H., Ren, G.: Almansi-type theorems in Clifford analysis. Math. Methods Appl. Sci. **25**, 1541–1552 (2002)
31. Sheldon, A., Paul, B., Wade, R.: Harmonic Function Theory. Graduate Texts in Mathematics, vol. 137. Springer, New York (2001)
32. Aronszajn, N., Creese, T.M., Lipkin, L.J.: Polyharmonic Functions. Oxford Mathematical Monographs, Oxford Science Publications. Oxford University Press, New York (1983)

Fueter Regularity and Slice Regularity: Meeting Points for Two Function Theories

Alessandro Perotti

Abstract We present some meeting points between two function theories, the Fueter theory of regular functions and the recent theory of quaternionic slice regular functions, which includes polynomials and power series with quaternionic coefficients. We show that every slice regular function coincides up to the first order with a unique regular function on the three-dimensional subset of reduced quaternions. We also characterize the regular functions so obtained. These relations have a higher dimensional counterpart between the theory of monogenic functions on Clifford algebras and the one of slice regular functions of a Clifford variable. We define a first order differential operator which extends the Dirac and Weyl operators to functions that can depend on all the coordinates of the algebra. The operator behaves well both w.r.t. monogenic functions and w.r.t. the powers of the (complete) Clifford variable. This last property relates the operator with the recent theory of slice monogenic and slice regular functions of a Clifford variable.

1 Introduction

The aim of this work is to illustrate some unexpected links between two function theories, one of which is well developed and dates back to the 1930's, while the other has been introduced recently but has seen rapid growth. The first is the Fueter theory of regular functions defined by means of the Cauchy-Riemann-Fueter differential operator, while the second is the theory of quaternionic slice regular functions, which comprises polynomials and power series with quaternionic coefficients on one side. These links have a higher dimensional counterpart between the theory of monogenic functions on Clifford algebras, defined in terms of the Dirac and Cauchy-Riemann operators, and the one of slice regular functions of a Clifford variable.

A. Perotti (✉)
Department of Mathematics, University of Trento, Via Sommarive 14, 38123 Povo (TN), Italy
e-mail: perotti@science.unitn.it

G. Gentili et al. (eds.), *Advances in Hypercomplex Analysis*, Springer INdAM Series 1,
DOI 10.1007/978-88-470-2445-8_6,

In order to obtain the promised relations between the two quaternionic function theories, we use a *modified* Cauchy-Riemann-Fueter operator $\mathcal{D}$, defined as:

$$\mathcal{D} = \frac{1}{2}\left(\frac{\partial}{\partial x_0} + i\frac{\partial}{\partial x_1} + j\frac{\partial}{\partial x_2} - k\frac{\partial}{\partial x_3}\right),$$

where x_0, x_1, x_2, x_3 are the real coordinates of an element $q = x_0 + ix_1 + jx_2 + kx_3$ of the quaternionic space $\mathbb{H}$ w.r.t. the basic elements i, j, k. We refer e.g. to [16, 21, 31] for some properties of this and other related differential operators on $\mathbb{H}$. The choice of $\mathcal{D}$ is justified by the fact that it behaves better than the standard Cauchy-Riemann-Fueter operator w.r.t. the powers of the quaternionic variable q. Moreover, the theory of functions defined by $\mathcal{D}$ has interesting relations with classes of holomorphic maps of two complex variables (see Sect. 2.1.1 for more details).

In general, the product of two Fueter regular functions is not Fueter regular. The same holds for functions in the kernel of $\mathcal{D}$. In particular, even if the identity function is regular, i.e. $\mathcal{D}(q) = 0$, the higher powers of q are not regular. Nevertheless, we are able to prove that for every positive power m, $\mathcal{D}(q^m)$ vanishes on the three-dimensional subset $\mathbb{H}_3 = \langle 1, i, j\rangle$ of reduced quaternions. This property extends to polynomials and convergent power series of the form $\sum_m q^m a_m$ and more generally to slice regular functions.

The concept of slice regularity for functions of one quaternionic, octonionic or Clifford variable has been introduced recently by Gentili and Struppa in [7, 8] and by Colombo, Sabadini and Struppa in [4] and further extended to real alternative *-algebras in [9, 10].

An application of the Cauchy-Kowalevski Theorem to the operator $\mathcal{D}$ assures that the restriction of any slice regular function to $\mathbb{H}_3$ has a unique regular (i.e. in the kernel of $\mathcal{D}$) extension to an open set. This extension gives an embedding of the space of slice regular functions into the space of regular functions.

The characterization of the image of this embedding is given by means of holomorphicity of the differentials w.r.t. the complex structures defined by left multiplication by imaginary reduced quaternions. It is based on a criterion for holomorphicity in the class of regular functions, which was proved in [22, 23] using the concept of the *energy quadric* of a function. The energy quadric is a positive semi-definite quadric, constructed by means of the Lichnerowicz homotopy invariants.

The second part of the paper is dedicated to the higher dimensional situation. We study some basic properties of a first order differential operator on the real Clifford algebra $\mathbb{R}_n$ of signature $(0, n)$ which generalizes the Weyl operator used in the theory of monogenic functions (for which we refer to [1, 3, 12]). While monogenic functions are usually defined on open subsets of the paravector space, the operator we consider acts on functions that can depend on all the coordinates of the algebra. This is similar to what happens in the quaternionic space $\mathbb{H} \simeq \mathbb{R}_2$, where the Cauchy-Riemann-Fueter operator acts on the whole space, not only on the reduced quaternions $\mathbb{H}_3$. Our starting point is the modified Cauchy-Riemann-Fueter operator $\mathcal{D}$. When written in the notation of the Clifford algebra $\mathbb{R}_2$, $\mathcal{D}$ becomes the operator

$$\mathcal{D}_2 = \frac{1}{2}\left(\frac{\partial}{\partial x_0} + e_1\frac{\partial}{\partial x_1} + e_2\frac{\partial}{\partial x_2} - e_{12}\frac{\partial}{\partial x_{12}}\right).$$

If $\mathcal{D}_1 = \frac{1}{2}(\frac{\partial}{\partial x_0} + e_1 \frac{\partial}{\partial x_1})$ and $\mathcal{D}_{1,2} = \frac{1}{2}(\frac{\partial}{\partial x_2} + e_1 \frac{\partial}{\partial x_{12}})$ are the one-variable Cauchy-Riemann operators w.r.t. the complex variables $z_1 = x_0 + e_1 x_1$, $z_2 = x_2 + e_1 x_{12}$, then $\mathcal{D}_2 = \mathcal{D}_1 + e_2 \mathcal{D}_{1,2}$. This observation suggests a recursive definition of a differential operator $\mathcal{D}_n$ on $\mathbb{R}_n$. Even if this definition of $\mathcal{D}_n$ is not symmetric w.r.t. the basis vectors, the operator we obtain *is* symmetric, and has the following explicit form:

$$\mathcal{D}_n = \frac{1}{2} \sum_K e_K^* \frac{\partial}{\partial x_K}$$

where $e_K^* = (-1)^{\frac{k(k-1)}{2}} e_K$ is obtained by applying to a basis element e_K the reversion anti-involution.

When restricted to functions of a paravector variable, $\mathcal{D}_n$ is equal (up to a factor 1/2) to the Weyl (cf. e.g. [3, Sect. 4.2]), or Cauchy-Riemann (as in [12, Sect. 5.3]) operator of $\mathbb{R}_n$. Therefore every $\mathbb{R}_n$-valued monogenic function defined on an open domain of the paravector subspace $\mathbb{R}^{n+1}$ of $\mathbb{R}_n$ is in the kernel of $\mathcal{D}_n$. Moreover, the identity function x of $\mathbb{R}_n$ is in the kernel of $\mathcal{D}_n$, while its restriction to the paravector variable is *not* monogenic. The operator $\mathcal{D}_n$ behaves well also w.r.t. powers of the (complete) Clifford variable x. We show that every power x^m is in the kernel of $\mathcal{D}_n$ when n is odd. For even n, the same property holds on the so-called *quadratic cone* of the algebra (cf. [9, 10]). These properties link the operators $\mathcal{D}_n$ to the recent theory of *slice monogenic* [4] and *slice regular* functions on $\mathbb{R}_n$ [9, 10].

Operators similar to $\mathcal{D}_n$ have already been considered in the literature (e.g. in [14, 28, 29]). However, it seems that the operators $\mathcal{D}_n$ are particularly well adapted to the theory of polynomials $\sum_m x^m a_m$ or more generally of slice regular functions on a Clifford algebra.

On the negative side, the operator $\mathcal{D}_n$ is not elliptic for $n > 2$ and its kernel is very large if we do not restrict the domains where functions are defined. In the last section, we focus on the case $n = 3$ and show a more strict relation of $\mathcal{D}_3$ with the Weyl operator. This suggests to consider a proper subspace of the kernel of $\mathcal{D}_3$, where the condition of *Cliffordian holomorphicity* [17] has a role. We get in this way the real analyticity in $\mathbb{R}_3$ and an integral representation formula on domains of polydisc type.

Some of the results of the present work have been presented in [26].

2 Fueter Regularity and Slice Regularity

We begin recalling some results of the Fueter theory of regular functions. We then introduce some definitions of the recent theory of quaternionic slice regular functions. Our approach uses a modification of the Fueter construction based on stem functions. We then present some meeting points between the two function theories. We show that every slice regular function coincides, up to the first order, with a unique regular function on the three-dimensional subset $\mathbb{H}_3 = \langle 1, i, j \rangle$ of reduced quaternions.

2.1 Fueter Regular Functions

We identify the space $\mathbb{C}^2$ with the set $\mathbb{H}$ of quaternions by means of the mapping that associates the pair $(z_1, z_2) = (x_0+ix_1, x_2+ix_3)$ with the quaternion $q = z_1+z_2j = x_0+ix_1+jx_2+kx_3 \in \mathbb{H}$. Given a bounded domain Ω in $\mathbb{H} \simeq \mathbb{C}^2$, a quaternionic function $f = f_1 + f_2 j$ of class C^1 on Ω will be called (left) *regular* on Ω if it is in the kernel of the *(modified) Cauchy-Riemann-Fueter operator*

$$\mathcal{D} = \frac{\partial}{\partial \bar{z}_1} + j\frac{\partial}{\partial \bar{z}_2} = \frac{1}{2}\left(\frac{\partial}{\partial x_0} + i\frac{\partial}{\partial x_1} + j\frac{\partial}{\partial x_2} - k\frac{\partial}{\partial x_3}\right) \quad \text{on } \Omega. \tag{1}$$

We will denote by $\mathcal{R}(\Omega)$ the real vector space of regular functions on Ω. The space $\mathcal{R}(\Omega)$ contains the identity mapping and every holomorphic mapping (f_1, f_2) on Ω (w.r.t. the standard complex structure) defines a regular function $f = f_1 + f_2 j$.

Given the decomposition in *real* components $f = f^0 + if^1 + jf^2 + kf^3$ of f, the operator $\mathcal{D}$ has the form:

$$\begin{aligned} 2\mathcal{D}(f) = {} & \frac{\partial f^0}{\partial x_0} - \frac{\partial f^1}{\partial x_1} - \frac{\partial f^2}{\partial x_2} + \frac{\partial f^3}{\partial x_3} + i\left(\frac{\partial f^1}{\partial x_0} + \frac{\partial f^0}{\partial x_1} + \frac{\partial f^3}{\partial x_2} + \frac{\partial f^2}{\partial x_3}\right) \\ & + j\left(\frac{\partial f^2}{\partial x_0} - \frac{\partial f^3}{\partial x_1} + \frac{\partial f^0}{\partial x_2} - \frac{\partial f^1}{\partial x_3}\right) + k\left(-\frac{\partial f^3}{\partial x_0} - \frac{\partial f^2}{\partial x_1} + \frac{\partial f^1}{\partial x_2} + \frac{\partial f^0}{\partial x_3}\right). \end{aligned}$$

We recall some properties of regular functions, for which we refer to the papers of Naser [19], Nōno [20], Sudbery [33], Shapiro and Vasilevski [31], Kravchenko and Shapiro [16]:

(i) Every regular function is harmonic: $\mathcal{D}\overline{\mathcal{D}} = \overline{\mathcal{D}}\mathcal{D} = \frac{1}{4}\Delta_4$, where

$$\overline{\mathcal{D}} = \frac{1}{2}\left(\frac{\partial}{\partial x_0} - i\frac{\partial}{\partial x_1} - j\frac{\partial}{\partial x_2} + k\frac{\partial}{\partial x_3}\right). \tag{2}$$

(ii) The space $\mathcal{R}(\Omega)$ of regular functions on Ω is a *right* $\mathbb{H}$-module with integral representation formulas.
(iii) Given the decomposition in *complex* components $f = f_1 + f_2 j$, f is regular if and only if $\frac{\partial f_1}{\partial \bar{z}_1} = \frac{\partial \overline{f_2}}{\partial z_2}$, $\frac{\partial f_1}{\partial \bar{z}_2} = -\frac{\partial \overline{f_2}}{\partial z_1}$.
(iv) The complex components f_1, f_2 are both holomorphic or both non-holomorphic.
(v) If Ω is pseudoconvex, every complex harmonic function is a complex component of a regular function on Ω.

The definition of regularity is equivalent to a notion introduced by Joyce [15] in the setting of hypercomplex manifolds. A hypercomplex structure on the manifold $\mathbb{H}$ is given by the complex structures J_1, J_2 on the tangent bundle $T\mathbb{H} \simeq \mathbb{H}$ defined by left multiplication by i and j. Let J_1^*, J_2^* be the dual structures on $T^*\mathbb{H} \simeq \mathbb{H}$ and set $J_3^* := J_1^* J_2^*$, which is equivalent to $J_3 = -J_1 J_2$. A function f is regular if and only if f is *q-holomorphic*, i.e. its differential $\mathrm{d}f$ satisfies the equation

$$\mathrm{d}f + iJ_1^*(\mathrm{d}f) + jJ_2^*(\mathrm{d}f) + kJ_3^*(\mathrm{d}f) = 0 \tag{3}$$

or, equivalently,

$$\mathrm{d}f^0 = J_1^*(\mathrm{d}f^1) + J_2^*(\mathrm{d}f^2) + J_3^*(\mathrm{d}f^3). \tag{4}$$

In complex components $f = f_1 + f_2 j$, we can rewrite the equations of regularity as

$$\overline{\partial} f_1 = J_2^*(\partial \overline{f}_2), \tag{5}$$

where $\partial = \sum_{i=1}^{2} \frac{\partial}{\partial z_i} \mathrm{d}z_i$ and $\overline{\partial} = \sum_{i=1}^{2} \frac{\partial}{\partial \bar{z}_i} \mathrm{d}\bar{z}_i$ are the Cauchy-Riemann differential operators on $\mathbb{C}^2$ w.r.t. the standard complex structure.

Remark 1 The original definition of regularity given by Fueter (cf. e.g. [12, 33]) considered the differential operator

$$\frac{\partial}{\partial x_0} + i\frac{\partial}{\partial x_1} + j\frac{\partial}{\partial x_2} + k\frac{\partial}{\partial x_3}, \tag{6}$$

which differs from $\mathcal{D}$ in the sign of the last derivative. If γ denotes the real reflection of $\mathbb{C}^2 \simeq \mathbb{R}^4$ defined by $\gamma(z_1, z_2) = (z_1, \bar{z}_2)$, then a C^1 function f is regular on the domain Ω if and only if $f \circ \gamma$ is Fueter–regular on $\gamma(\Omega) = \gamma^{-1}(\Omega)$, i.e. it satisfies the differential equation

$$\left(\frac{\partial}{\partial x_0} + i\frac{\partial}{\partial x_1} + j\frac{\partial}{\partial x_2} + k\frac{\partial}{\partial x_3}\right)(f \circ \gamma) = 0 \tag{7}$$

on $\gamma^{-1}(\Omega)$. The reflection γ has an algebraic interpretation. It can be seen as the *reversion* anti-involution $q \mapsto q^*$ of the Clifford algebra $\mathbb{R}_2 \simeq \mathbb{H}$, defined by

$$q^* = (x_0 + ix_1 + jx_2 + kx_3)^* = x_0 + ix_1 + jx_2 - kx_3. \tag{8}$$

2.1.1 Holomorphic Functions w.r.t. a Complex Structure J_p

Let $J_p = p_1 J_1 + p_2 J_2 + p_3 J_3$ be the *orthogonal complex structure* on $\mathbb{H}$ defined by a quaternion $p = p_1 i + p_2 j + p_3 k$ in the sphere $\mathbb{S} = \{p \in \mathbb{H} \mid p^2 = -1\} \simeq S^2$ of quaternionic imaginary units. In particular, J_1 is the standard complex structure of $\mathbb{C}^2 \simeq \mathbb{H}$. Let $\mathbb{C}_p = \langle 1, p \rangle$ be the complex plane spanned by 1 and p and let L_p be the complex structure defined on $T^*\mathbb{C}_p \simeq \mathbb{C}_p$ by left multiplication by p. Then $L_p = J_{p^*}$, where $p^* = p_1 i + p_2 j - p_3 k$.

Let $\mathrm{Hol}_p(\Omega, \mathbb{H})$ be the space of holomorphic maps between the (almost) complex manifolds (Ω, J_p) and $(\mathbb{H}, L_p)$:

$$\mathrm{Hol}_p(\Omega, \mathbb{H}) = \{f : \Omega \to \mathbb{H} \mid \overline{\partial}_p f = 0 \text{ on } \Omega\} = \mathrm{Ker}\, \overline{\partial}_p, \tag{9}$$

where $\overline{\partial}_p$ is the Cauchy-Riemann operator

$$\overline{\partial}_p = \frac{1}{2}\left(\mathrm{d} + pJ_p^* \circ \mathrm{d}\right). \tag{10}$$

These functions will be called *J_p-holomorphic maps* on Ω. For any positive orthonormal basis $\{1, p, q, pq\}$ of $\mathbb{H}$ defined by a orthogonal pair $p, q \in \mathbb{S}^2$, let $f = f_1 + f_2 q$ be the decomposition of f with respect to the orthogonal sum

$$\mathbb{H} = \mathbb{C}_p \oplus (\mathbb{C}_p)q. \tag{11}$$

Let $f_1 = f^0 + pf^1$, $f_2 = f^2 + pf^3$, with f^0, f^1, f^2, f^3 the real components of f w.r.t. the basis $\{1, p, q, pq\}$. Then the equations of regularity can be rewritten in complex form as

$$\overline{\partial}_p f_1 = J_q^*(\partial_p \overline{f}_2), \tag{12}$$

where $\overline{f}_2 = f^2 - pf^3$ and $\partial_p = \frac{1}{2}(\mathrm{d} - pJ_p^* \circ \mathrm{d})$. Therefore every $f \in \mathrm{Hol}_p(\Omega, \mathbb{H})$ is a regular function on Ω.

Remark 2 We refer to [22, 25] for the following properties of J_p-holomorphic maps.

(i) The identity map belongs to the spaces $\mathrm{Hol}_i(\mathbb{H}, \mathbb{H})$ and $\mathrm{Hol}_j(\mathbb{H}, \mathbb{H})$ but not to $\mathrm{Hol}_k(\mathbb{H}, \mathbb{H})$.
(ii) For every $p \in \mathbb{S}^2$, the spaces $\mathrm{Hol}_{-p}(\Omega, \mathbb{H})$ and $\mathrm{Hol}_p(\Omega, \mathbb{H})$ coincide.
(iii) Every $\mathbb{C}_p$-valued regular function is a J_p-holomorphic function.
(iv) If $f \in \mathrm{Hol}_p(\Omega, \mathbb{H}) \cap \mathrm{Hol}_q(\Omega, \mathbb{H})$, with $p \neq \pm q$, then $f \in \mathrm{Hol}_r(\Omega, \mathbb{H})$ for every $r = \frac{\alpha p + \beta q}{\|\alpha p + \beta q\|}$ $(\alpha, \beta \in \mathbb{R})$ in the circle of $\mathbb{S}^2$ generated by p and q.

In [23] was proved that on every domain Ω there exist regular functions that are not J_p-holomorphic for any p. For example, the linear function $f = z_1 + z_2 + \bar{z}_1 + (z_1 + z_2 + \bar{z}_2)j$ is regular on $\mathbb{H}$, but not holomorphic. The criterion for holomorphicity is based on an energy–minimizing property of holomorphic maps (see Sect. 2.4.4 for definitions and properties of the *energy quadric* of a quaternionic function f).

We can obtain regular functions also when considering non-constant (almost) complex structures. If $p = p(z) \in \mathbb{S}$ varies smoothly with $z \in \Omega$, the almost complex structures $J_{p(z)}$ and $L_{p(z)}$ are not constant, i.e. not compatible with the hyperkähler structure of $\mathbb{H}$. Note that in this case the structures are not necessarily integrable. Let $f \in C^1(\Omega)$. We shall say that p is *f-equivariant* if $f(z) = f(z')$ implies $p(z) = p(z')$ for $z, z' \in \Omega$. This property allows to define $\tilde{p} : f(\Omega) \to \mathbb{S}^2$ such that $\tilde{p} \circ f = p$ on Ω. In [22] was proved that $J_{p(z)}$-holomorphic maps $f : (\Omega, J_{p(z)}) \to (\mathbb{H}, L_{p(f(z))})$ give rise to regular functions:

Proposition 1 ([22, Proposition 1]) *If $f \in C^1(\Omega)$ satisfies the equation*

$$\overline{\partial}_{p(z)} f = \frac{1}{2}\big[\mathrm{d}f(z) + p(z)J_{p(z)}^* \circ \mathrm{d}f(z)\big] = 0 \tag{13}$$

at every $z \in \Omega$, then f is a regular function on Ω. If, moreover, the structure p is f-equivariant and $\tilde{p}$ admits a continuous extension to an open set $U \supseteq f(\Omega)$, then f is a (pseudo)holomorphic map from (Ω, J_p) to $(U, L_{\tilde{p}})$.

For example, the function $f(z) = \bar{z}_1 + z_2^2 + \bar{z}_2 j$ is regular on $\mathbb{H}$. On the set $\Omega = \mathbb{H} \setminus \{z_2 = 0\}$ f is holomorphic w.r.t. the almost complex structure $J_{p(z)}$, where

$$p(z) = \frac{1}{\sqrt{|z_2|^2 + |z_2|^4}}\big(|z_2|^2 i - (\mathrm{Im}\, z_2)j - (\mathrm{Re}\, z_2)k\big). \tag{14}$$

2.2 Fueter Construction

In 1934, Rudolf Fueter [6] proposed a simple method, which is now known as Fueter's Theorem, to generate quaternionic regular functions by means of complex holomorphic functions. Given a holomorphic "stem function"

$$F(z) = u(\alpha, \beta) + \mathrm{i}\, v(\alpha, \beta) \quad (z = \alpha + \mathrm{i}\beta \text{ complex}, \ u, v \text{ real-valued})$$

defined in the upper complex half-plane, real-valued when restricted to the real line, the formula:

$$f(q) := u\big(x_0, |\mathrm{Im}(q)|\big) + \frac{\mathrm{Im}(q)}{|\mathrm{Im}(q)|} v\big(x_0, |\mathrm{Im}(q)|\big) \tag{15}$$

(with $q = x_0 + x_1 i + x_2 j + x_3 k \in \mathbb{H}$, $\mathrm{Im}(q) = x_1 i + x_2 j + x_3 k$) defines a *radially holomorphic* function on $\mathbb{H}$, whose Laplacian Δf is Fueter regular (now called the *Fueter transform* of F). Fueter's construction was later extended to higher dimensions by Sce [30], Qian [27] and Sommen [32] in the setting of octonionic and Clifford analysis. Fueter's Theorem and its generalizations provides a link between slice regular functions and Fueter regular (resp. monogenic) functions. In the next sections, we will present a relation of a different kind between these function theories.

2.3 Quaternionic Slice Regular Functions

A modification of the Fueter construction can be applied to give a new approach (cf. [9, 10]) to the concept of "slice regularity" for functions of one quaternionic, octonionic or Clifford variable, which has been recently introduced by Gentili and Struppa in [7, 8] and by Colombo, Sabadini and Struppa in [4]. We start with a holomorphic function $F(z)$ with *quaternionic*-valued components u, v:

$$F(z) = u(\alpha, \beta) + \mathrm{i}\, v(\alpha, \beta) \quad (z = \alpha + \mathrm{i}\beta \text{ complex}, \ u, v \ \mathbb{H}\text{-valued})$$

defined on a subset of the upper complex half-plane, real-valued on $\mathbb{R}$. Formula (15) defines then a *slice regular* (or *Cullen regular*) [7, 8] function on an open subset of the quaternionic space. We now make more precise this idea.

Let $q^c = x_0 - ix_1 - jx_2 - kx_4$ denote the *quaternionic conjugation*. Let $\mathbb{H}_\mathbb{C} = \mathbb{H} \otimes_\mathbb{R} \mathbb{C}$ be the complexified quaternion algebra. We will use the representation

$$\mathbb{H}_\mathbb{C} = \{w = x + \mathrm{i}y \mid x, y \in \mathbb{H}\}. \tag{16}$$

$\mathbb{H}_\mathbb{C}$ is a complex algebra with unity w.r.t. the product defined as follows:

$$(x + \mathrm{i}y)\big(x' + \mathrm{i}y'\big) = xx' - yy' + \mathrm{i}\big(xy' + yx'\big). \tag{17}$$

In $\mathbb{H}_\mathbb{C}$ two commuting operations are defined: the *anti-involution*

$$w \mapsto w^c = (x + \mathrm{i}y)^c = x^c + \mathrm{i}y^c \tag{18}$$

and the *complex conjugation*

$$w \mapsto \overline{w} = \overline{x + \mathrm{i}y} = x - \mathrm{i}y. \tag{19}$$

Definition 1 Let $D \subseteq \mathbb{C}$ be an open subset. If a function $F : D \to \mathbb{H}_\mathbb{C}$ is *complex intrinsic*, i.e. it satisfies the condition $F(\overline{z}) = \overline{F(z)}$ for every $z \in D$ such that $\overline{z} \in D$, then F is called a *stem function* on D.

Remark 3

(i) In the preceding definition, there is no restriction to assume that D is symmetric w.r.t. the real axis, i.e. $D = \mathrm{conj}(D) := \{z \in \mathbb{C} \mid \overline{z} \in D\}$.
(ii) A function F is a stem function if and only if the $\mathbb{H}$-valued components F_1, F_2 of $F = F_1 + \mathrm{i}F_2$ form an *even-odd pair* w.r.t. the imaginary part of z:
$$F_1(\overline{z}) = F_1(z), \qquad F_2(\overline{z}) = -F_2(z) \quad \text{for every } z \in D. \tag{20}$$
(iii) By means of a real basis $\mathcal{B}$ of $\mathbb{H}$, F can be identified with a complex intrinsic curve $F^\mathcal{B}$ in $\mathbb{C}^4$.

Given an open subset D of $\mathbb{C}$, let Ω_D be the open subset of $\mathbb{H}$ obtained by the action on D of the square roots of -1:
$$\Omega_D := \{q = \alpha + \beta p \in \mathbb{C}_p \mid \alpha, \beta \in \mathbb{R},\ \alpha + i\beta \in D,\ p \in \mathbb{S}\}. \tag{21}$$
Sets of this type will be called *circular sets* in $\mathbb{H}$.

Definition 2 Any stem function $F : D \to \mathbb{H}_\mathbb{C}$ induces a *left slice function* $f = \mathcal{I}(F) : \Omega_D \to \mathbb{H}$. If $q = \alpha + \beta p \in D_p := \Omega_D \cap \mathbb{C}_p$, with $p \in \mathbb{S}$, we set
$$f(q) := F_1(z) + pF_2(z) \quad (z = \alpha + i\beta \in D).$$

Note that if $q = x_0 + ix_1 + jx_2 + kx_3 = x_0 + \mathrm{Im}(q)$ and $\mathrm{Im}(q) \neq 0$, then $p = \pm\,\mathrm{Im}(q)/|\mathrm{Im}(q)|$. If $\mathrm{Im}(q) = 0$, then every choice of $p \in \mathbb{S}$ can be done. The complex intrinsicity of the stem function F assures that the definition of f is well posed.

There is an analogous definition for *right* slice functions when p is placed on the right of $F_2(z)$. In what follows, the term *slice functions* will always mean left slice functions.

We will denote the real vector space of (left) slice functions on Ω_D by $\mathcal{S}(\Omega_D)$. We will denote by $\mathcal{S}^1(\Omega_D) := \{f = \mathcal{I}(F) \in \mathcal{S}(\Omega_D) \mid F \in C^1(D)\}$ the real vector space of slice functions with stem function of class C^1. It can be shown (cf. [10]) that every $f \in \mathcal{S}^1(\Omega_D)$ is of class C^1 on Ω_D.

Definition 3 Let $f = \mathcal{I}(F) \in \mathcal{S}^1(\Omega_D)$. We set
$$\frac{\partial f}{\partial q} := \mathcal{I}\left(\frac{\partial F}{\partial z}\right), \qquad \frac{\partial f}{\partial q^c} := \mathcal{I}\left(\frac{\partial F}{\partial \overline{z}}\right).$$
These functions are continuous slice functions on Ω_D.

We now introduce slice regularity. Left multiplication by i defines a complex structure on $\mathbb{H}_\mathbb{C}$. With respect to this structure, a C^1 function $F = F_1 + \mathrm{i}F_2 :$

$D \to \mathbb{H}_{\mathbb{C}}$ is holomorphic if and only if its components F_1, F_2 satisfy the Cauchy-Riemann equations:

$$\frac{\partial F_1}{\partial \alpha} = \frac{\partial F_2}{\partial \beta}, \qquad \frac{\partial F_1}{\partial \beta} = -\frac{\partial F_2}{\partial \alpha}, \quad \text{i.e.} \quad \frac{\partial F}{\partial \bar{z}} = 0. \tag{22}$$

Definition 4 A (left) slice function $f \in \mathcal{S}^1(\Omega_D)$ is *(left) slice regular* if its stem function F is holomorphic. We will denote the real vector space of slice regular functions on Ω_D by $\mathcal{SR}(\Omega_D) := \{f \in \mathcal{S}^1(\Omega_D) \mid f = \mathcal{I}(F),\ \ F : D \to \mathbb{H}_{\mathbb{C}}$ holomorphic$\}$.

Polynomials $p(q) = \sum_{m=0}^{d} q^m a_m = \mathcal{I}(\sum_{m=0}^{d} z^m a_m)$ with right quaternionic coefficients can be considered as slice regular functions on $\mathbb{H}$. More generally, every convergent *power series* $\sum_m q^m a_m$ is a slice regular function on an open ball of $\mathbb{H}$ centered in the origin with (possibly infinite) positive radius.

Proposition 2 ([10, Proposition 8]) *Let* $f = \mathcal{I}(F) \in \mathcal{S}^1(\Omega_D)$. *Then* f *is slice regular on* Ω_D *if and only if for every* $p \in \mathbb{S}$ *the restriction* $f_p := f_{|\mathbb{C}_p \cap \Omega_D} : D_p = \mathbb{C}_p \cap \Omega_D \to \mathbb{H}$ *is holomorphic with respect to the complex structure* J_p *defined by left multiplication by* p.

Proposition 2 means that if D intersects the real axis, f is slice regular on Ω_D if and only if it is *Cullen regular* in the sense introduced by Gentili and Struppa in [7, 8].

2.4 A 3D-Meeting Point for Two Function Theories

We start by computing the values of $\mathcal{D}(q^m)$. The crucial observation is that these values vanishes on the subspace of reduced quaternions. This property allows to extend the restriction to $\mathbb{H}_3$ of a slice regular function to a regular function.

2.4.1 Computation of $\mathcal{D}(q^m)$

In general, the product of two Fueter regular functions is not Fueter regular. The same holds for functions in the kernel of the modified Cauchy-Riemann-Fueter operator $\mathcal{D}$. In particular, even if the identity function is regular, i.e. $\mathcal{D}(q) = 0$, the higher powers of q are not regular. Nevertheless, we can show that $\mathcal{D}(q^m)$ vanishes on a three-dimensional subset of $\mathbb{H}$ for every positive power m.

Proposition 3 *Let* $\mathcal{D}$ *be the Cauchy-Riemann-Fueter operator of Sect.* 2.1. *Given two functions* f, g *of class* C^1, *the operator* $\mathcal{D}$ *satisfies the following product formula*:

$$\mathcal{D}(fg) = \mathcal{D}(f)g + \frac{1}{2}\left(f\frac{\partial g}{\partial x_0} + if\frac{\partial g}{\partial x_1} + j\,f\frac{\partial g}{\partial x_2} - kf\frac{\partial g}{\partial x_3}\right). \tag{23}$$

As a consequence, we get the following power formula for the (modified) Cauchy-Riemann-Fueter operator:

$$\mathcal{D}(q^m) = \mathcal{D}(q^{m-1})q + q^{m-1} - (q^{m-1})^* \tag{24}$$

where $q^* = x_0 + ix_1 + jx_2 - kx_3$ *is obtained applying the reversion anti-involution to* q *(cf. Remark* 1*).*

Proof The product formula (23) follows immediately from the definition of $\mathcal{D}$. When applied to $q^m = q^{m-1}q$ it gives:

$$\mathcal{D}(q^m) = \mathcal{D}(q^{m-1})q + \frac{1}{2}(q^{m-1} + iq^{m-1}i + jq^{m-1}j - kq^{m-1}k). \tag{25}$$

For every quaternion $p = x_0 + ix_1 + jx_2 + kx_3$, the sum $p + ipi + jpj - kpk$ is equal to $4kx_3 = 2(p - p^*)$, from which (24) follows. □

Let $\mathbb{H}_3 = \langle 1, i, j \rangle$ be the real vector subspace of *reduced quaternions*. It is well known (cf. e.g. [13]) that the powers of a reduced quaternion are still reduced quaternions. This follows easily from the fact that reduced quaternions are characterized by the condition $q = q^*$. Therefore, if $p, q \in \mathbb{H}_3$, $(pq)^* = q^* p^* = qp$, and then pq is still reduced if and only if p and q commute, i.e. $q \in \mathbb{C}_p$.

Corollary 1 *Let* m *be a positive integer. Then* $\mathcal{D}(q^m)$ *vanishes on* $\mathbb{H}_3$*.*

Proof Since the power function $q \mapsto q^m$ maps $\mathbb{H}_3$ into $\mathbb{H}_3$, $q^{m-1} - (q^{m-1})^*$ vanishes on $\mathbb{H}_3$ for every $m \geq 1$. We conclude by induction on m starting from $\mathcal{D}(q) = 0$ and using (24). □

Corollary 2 *For every convergent power series* $f(q) = \sum_m q^m a_m$ *with quaternionic coefficients,* $\mathcal{D}(f)$ *vanishes on the intersection of the ball of convergence with the real vector space* $\mathbb{H}_3 \simeq \mathbb{R}^3$ *of reduced quaternions.* □

Corollary 3 *Assume that* Ω_D *is connected and* $\Omega_D \cap \mathbb{R} \neq \emptyset$*. Let* $f \in \mathcal{SR}(\Omega_D)$*. Then* $\mathcal{D}(f)$ *vanishes at every point of* $\widetilde{\Omega}_D := \Omega_D \cap \mathbb{H}_3$*.*

Proof Every slice regular function $f \in \mathcal{SR}(\Omega_D)$ has convergent power series expansions $\sum_m (q - r)^m a_m$ centered at real points r of Ω_D [7, 8]. Since every power $(q - r)^m$ is a polynomial in q, the thesis follows from Corollary 1. □

We observe that the previous results are not true for the standard (not modified) Cauchy-Riemann-Fueter operator.

2.4.2 Regular Extension of Slice-Regular Functions

Let $f \in \mathcal{SR}(\Omega_D)$. Then f is real analytic on Ω_D (cf. [10, Proposition 7]). We can apply the Cauchy-Kowalevski Theorem (see for example [5]) to the initial values problem

$$\begin{cases} \mathcal{D}(g) = -\mathcal{D}(f) & \text{on } \Omega_D \\ g = 0 & \text{on } \widetilde{\Omega}_D = \Omega_D \cap \mathbb{H}_3 \end{cases} \tag{26}$$

and obtain a real analytic solution in the neighborhood of every point of the intersection of Ω_D with the hyperplane $\mathbb{H}_3$. By taking the union of all these neighborhoods, we get the existence of a solution of problem (26) on a open set $\Omega' \subset \mathbb{H}$ such that $\Omega' \cap \mathbb{H}_3 = \widetilde{\Omega}_D$.

Since $\mathcal{D}(f) = 0$ and $g = 0$ on $\widetilde{\Omega}_D$, also the normal derivative

$$\begin{aligned} \frac{\partial g}{\partial x_3} &= 2k\mathcal{D}(g) + k\left(\frac{\partial g}{\partial x_0} + i\frac{\partial g}{\partial x_1} + j\frac{\partial g}{\partial x_2}\right) \\ &= -2k\mathcal{D}(f) + k\left(\frac{\partial g}{\partial x_0} + i\frac{\partial g}{\partial x_1} + j\frac{\partial g}{\partial x_2}\right) \end{aligned} \tag{27}$$

vanishes on $\widetilde{\Omega}_D$. Therefore $g = 0$, $dg = 0$ on $\widetilde{\Omega}_D$. This implies that the slice regular function $f \in \mathcal{SR}(\Omega_D)$ and the regular function $f + g \in \mathcal{R}(\Omega') \subset \text{Ker}(\mathcal{D})$ coincide up to the first order on the three-dimensional set $\widetilde{\Omega}_D$. We summarize the result in the following statement.

Proposition 4 *Assume that Ω_D is connected and $\Omega_D \cap \mathbb{R} \neq \emptyset$. Let $f \in \mathcal{SR}(\Omega_D)$. Then there exists an open (relative to $\mathbb{H}$) neighborhood Ω' of $\Omega_D \cap \mathbb{H}_3$ and a regular function $\tilde{f} \in \mathcal{R}(\Omega')$ such that $f = \tilde{f}$ on $\Omega' \cap \mathbb{H}_3 = \Omega_D \cap \mathbb{H}_3$ up to the first order.* □

For polynomials $f(q) = \sum_m q^m a_m$, the solution to problem (26) can be obtained explicitly in a finite number of steps by means of one-variable integrations w.r.t. the normal coordinate x_3. Consider the approximate problems:

$$\begin{cases} \dfrac{\partial g^{(h+1)}}{\partial x_3} = -2k\mathcal{D}\left(f + \displaystyle\sum_{l=1}^{h} g^{(l)}\right) & \text{on } \mathbb{H} \\ g^{(h+1)} = 0 & \text{on } \mathbb{H}_3 \end{cases} \tag{28}$$

for $h = 1, \ldots, \deg(f)$, starting with the function $g^{(1)} \equiv 0$. We are looking for polynomial solutions $g^{(h)}$ with $\deg(g^{(h)}) \leq h$.

At the first step, since x_3 divides the polynomial $\mathcal{D}(f)$ (from Corollary 1), x_3^2 divides the first solution $g^{(2)}$ of (28). Therefore x_3^2 divides also $\mathcal{D}(f + g^{(2)})$, since

$$2\mathcal{D}(f + g^{(2)}) = k\frac{\partial g^{(2)}}{\partial x_3} + 2\mathcal{D}(g^{(2)}) = \frac{\partial g^{(2)}}{\partial x_0} + i\frac{\partial g^{(2)}}{\partial x_1} + j\frac{\partial g^{(2)}}{\partial x_2} \tag{29}$$

and $g^{(2)} = 0$ on $\mathbb{H}_3$. By induction on h, we get by the same reasoning that the power x_3^h divides $g^{(h)}$ and $\mathcal{D}(f + \sum_{l=1}^{h} g^{(l)})$ for every $h = 1, \ldots, \deg(f)$. At the last step, we get that $x_3^{\deg(f)}$ divides $\mathcal{D}(f+\sum_{i=1}^{\deg(f)} g^{(i)})$, but then it has to be $\mathcal{D}(f+\sum_{i=1}^{\deg(f)} g^{(i)})=0$ for degree reasons. Observe that x_3^2 divides every partial solution $g^{(h)}$.

The polynomial $\tilde{f} := f+\sum_{i=1}^{\deg(f)} g^{(i)}$ is regular on the whole space, has the same degree as f, and coincides with f up to first order on $\mathbb{H}_3$.

As an illustration of this procedure, we compute the extension $\tilde{f}$ for the first three powers of q:

(i) $f(q)=q$ is regular, so $\tilde{f}=f$.
(ii) $f(q)=q^2$ has regular extension $\tilde{f}=q^2+2x_3^2$.
(iii) $f(q)=q^3$ has regular extension $\tilde{f}=q^3+x_3^2(6x_0+2x_1i+2x_2j+\frac{2}{3}x_3k)$.

2.4.3 The Embedding $\mathcal{SR}\hookrightarrow\mathcal{R}$

Definition 5 Assume that Ω_D is connected and $\Omega_D\cap\mathbb{R}\neq\emptyset$. Given $f\in\mathcal{SR}(\Omega_D)$, denote by $\mathrm{Reg}(f)$ the unique regular function defined on a maximal domain and satisfying

$$f=\mathrm{Reg}(f),\qquad \mathrm{d}f=\mathrm{d}\,\mathrm{Reg}(f)\quad\text{at every point of } \Omega_D\cap\mathbb{H}_3.$$

Uniqueness follows from the identity principle for regular functions. Let

$$\widetilde{\mathcal{R}}(\Omega_D):=\left\{g\in\mathcal{R}(\Omega')\;\middle|\;\Omega' \text{ open and connected in } \mathbb{H} \text{ s.t. } \Omega'\cap\mathbb{H}_3=\Omega_D\cap\mathbb{H}_3\right\}.$$

In the space $\widetilde{\mathcal{R}}(\Omega_D)$ we identify two functions if they coincide on the intersection of the domains of definition. The mapping $f\mapsto\mathrm{Reg}(f)$ is an injective (right) $\mathbb{H}$-linear operator, giving an embedding $\mathcal{SR}(\Omega_D)\hookrightarrow\widetilde{\mathcal{R}}(\Omega_D)$.

2.4.4 Characterization of $\mathrm{Reg}(\mathcal{SR}(\Omega_D))$ in $\widetilde{\mathcal{R}}(\Omega_D)$

Assume $\Omega_D\cap\mathbb{R}\neq\emptyset$. Let $f\in\mathcal{SR}(\Omega_D)$ be slice regular and $x\in\mathbb{H}_3\setminus\mathbb{R}$, let $p:=\mathrm{Im}(x)/|\mathrm{Im}(x)|\in\mathbb{S}$. From Proposition 2 we get that the restriction

$$f_p:=f_{|\mathbb{C}_p\cap\Omega_D}:\mathbb{C}_p\cap\Omega_D\to\mathbb{H}$$

is holomorphic with respect to the complex structure J_p defined by left multiplication by p. Moreover, at every $x\in\widetilde{\Omega}_D=\Omega_D\cap\mathbb{H}_3$, because of Proposition 4 the differential map

$$\mathrm{d}f_x=\mathrm{d}\,\mathrm{Reg}(f)_x$$

is a regular linear function on $\mathbb{H}$. These two properties can be strengthened in the sense explained by the next statement.

Theorem 1 *If $f\in\mathcal{SR}(\Omega_D)$ and $x\in\widetilde{\Omega}_D\setminus\mathbb{R}=(\Omega_D\cap\mathbb{H}_3)\setminus\mathbb{R}$, then the differential map $\mathrm{d}f_x$ belongs to the space $\mathrm{Hol}_{J_p}(\mathbb{H},\mathbb{H})\subset\mathcal{R}(\mathbb{H})$, where $p:=\mathrm{Im}(x)/|\mathrm{Im}(x)|$, i.e. the linear map*

$$\mathrm{d}f_x:(\mathbb{H},J_p)\to(\mathbb{H},J_p)$$

is holomorphic.

The proof of Theorem 1 is based on a criterion for holomorphicity in the space $\mathcal{R}(\Omega_D)$, which was proved in [22, 23] using the concept of the *energy quadric* of a function. The energy quadric of a regular function f is a positive semi-definite quadric, constructed by means of the Lichnerowicz homotopy invariants.

We recall some definitions and results from [22, 23]. The *energy density* of a map $f : \Omega \to \mathbb{H}$, of class $C^1(\Omega)$, w.r.t. the Euclidean metric, is the function

$$\mathcal{E}(f) := \frac{1}{2}\|\mathrm{d}f\|^2 = \frac{1}{2}\operatorname{tr}\big(Jac(f)\overline{Jac(f)}^T\big),$$

where $Jac(f)$ is the Jacobian matrix of f. Assume Ω relatively compact. The *energy* of $f \in C^1(\overline{\Omega})$ on Ω is the integral defined by

$$\mathcal{E}_\Omega(f) := \int_\Omega \mathcal{E}(f)\,\mathrm{d}V.$$

Let $A = (a_{\alpha\beta})$ be the 3×3 matrix with entries the real functions

$$a_{\alpha\beta} = -\langle J_\alpha, f^* L_{i_\beta}\rangle, \quad \text{where } (i_1, i_2, i_3) = (i, j, k).$$

(These numbers are the analogues of the Lichnerowicz invariants (cf. [18] and [2].) For $f \in C^1(\overline{\Omega})$, we set

$$A_\Omega := \int_\Omega A\,\mathrm{d}V \quad \text{and} \quad M_\Omega := \frac{1}{2}\big((\operatorname{tr} A_\Omega)I_3 - A_\Omega\big),$$

where I_3 denotes the identity matrix.

Theorem 2 ([23]) *If $f \in C^1(\overline{\Omega})$ is regular on Ω, then it minimizes energy in its homotopy class (relative to $\partial\Omega$).*

Theorem 3 ([23]) *Let $f \in C^1(\overline{\Omega})$. The following facts hold:*

(i) *f is regular on Ω if and only if $\mathcal{E}_\Omega(f) = \operatorname{tr} M_\Omega$.*
(ii) *If $f \in \mathcal{R}(\Omega)$, then M_Ω is symmetric and positive semidefinite.*
(iii) *If $f \in \mathcal{R}(\Omega)$, then f belongs to some space $\mathrm{Hol}_p(\Omega, \mathbb{H})$ (for a constant structure J_p) if and only if $\det M_\Omega = 0$. More precisely, $X_p = (p_1, p_2, p_3)$ is a unit vector in the kernel of M_Ω if and only if $f \in \mathrm{Hol}_p(\Omega, \mathbb{H})$.*

The criterion of holomorphicity holds also pointwise: let Ω be connected and $f \in C^1(\Omega)$. Consider the matrix of real functions on Ω:

$$M := \frac{1}{2}\big((\operatorname{tr} A)I_3 - A\big).$$

Theorem 4 ([22]) *Let $f \in C^1(\Omega)$. The following facts hold:*

(i) *f is regular on Ω if and only if $\mathcal{E}(f) = \operatorname{tr} M$ at every point $z \in \Omega$.*
(ii) *If $f \in \mathcal{R}(\Omega)$, then M is a 3×3 symmetric and positive semidefinite matrix.*
(iii) *If $f \in \mathcal{R}(\Omega)$, then $\det M = 0$ on Ω if and only if there exists an open, dense subset $\Omega' \subseteq \Omega$ such that f is a (pseudo)holomorphic map from $(\Omega', J_{p(z)})$ to $(\mathbb{H}, L_{p(f(z))})$ for some $p(z)$.*

Now we come to the *proof of Theorem* 1. To this end, we compute the energy quadric of f at $x \in \Omega_D \cap \mathbb{H}_3$. Since f coincides with $\mathrm{Reg}(f)$ up to first order on $\Omega_D \cap \mathbb{H}_3$, from Theorem 4 we get that $\mathcal{E}(f) = \mathrm{tr}\, M(f)$ on $\Omega_D \cap \mathbb{H}_3$ and that $M(f)$ is positive semidefinite at every point $x \in \Omega_D \cap \mathbb{H}_3$. Let us denote by $\pi_1(f) = f_1$ and $\pi_2(f) = f_2$ the complex components of f. From [22] we get the following expression of the energy quadric $M(f)$ at $x \in \Omega_D \cap \mathbb{H}_3$ in terms of f_1 and f_2 and their complex derivatives:

$$M(f) = \begin{bmatrix} 2|c|^2 & \mathrm{Im}\langle c, a-b\rangle & \mathrm{Re}\langle c, a+b\rangle \\ \mathrm{Im}\langle c, a-b\rangle & \frac{1}{2}|a-b|^2 & -\mathrm{Im}\langle a, b\rangle \\ \mathrm{Re}\langle c, a+b\rangle & -\mathrm{Im}\langle a, b\rangle & \frac{1}{2}|a+b|^2 \end{bmatrix}, \tag{30}$$

where

$$a = \left(\frac{\partial f_1}{\partial z_1}, \frac{\partial f_2}{\partial z_1}\right), \qquad b = \left(\frac{\partial \bar{f}_2}{\partial \bar{z}_2}, -\frac{\partial \bar{f}_1}{\partial \bar{z}_2}\right), \qquad c = \left(\frac{\partial \bar{f}_2}{\partial z_1}, -\frac{\partial \bar{f}_1}{\partial z_1}\right)$$

are all computed at $x \in \Omega_D \cap \mathbb{H}_3$. We now show that the vector $(x_1, x_2, 0)$ belongs to the null space of the matrix $M(q^m)$. If $(x_1, x_2, 0) \neq (0, 0, 0)$, it is possible to find a similarity of the space $\langle i, j, k\rangle \simeq \mathbb{R}^3$, with rotational component induced by a reduced quaternion $a \in \mathbb{H}_3$, which sends $(x_1, x_2, 0)$ in $(1, 0, 0)$. The transformation property of the energy quadric w.r.t. rotations (see [24, Theorem 4]) implies that $(x_1, x_2, 0) \in \mathrm{Ker}(M(q^m))$ at $x = x_0 + ix_1 + jx_2$ if and only if $(1, 0, 0) \in \mathrm{Ker}(M(q^m))$ at $x_0 + i$.

In view of (30), in order to show that the vector $(1, 0, 0)$ belongs to the null space of $M(q^m)$ at $x_0 + i$ it suffices to prove that $\bar{c} = (\frac{\partial \pi_2(q^m)}{\partial \bar{z}_1}, -\frac{\partial \pi_1(q^m)}{\partial \bar{z}_1})$ vanishes at $x_0 + i$. This is a consequence of the following lemma.

Lemma 1 *For every positive integer m, z_2 divides on the left the partial derivative $\frac{\partial q^m}{\partial \bar{z}_1}$, i.e. $\frac{\partial q^m}{\partial \bar{z}_1} = z_2 g$ for a quaternionic function g.*

Proof For $m > 1$ the following product formula can be easily obtained:

$$\frac{\partial q^m}{\partial \bar{z}_1} = \frac{\partial q^{m-1}}{\partial \bar{z}_1} q + \frac{1}{2}\left(q^{m-1} + i q^{m-1} i\right) = \frac{\partial q^{m-1}}{\partial \bar{z}_1} q + \pi_2\left(q^{m-1}\right) j. \tag{31}$$

Using (31) and the equation

$$\begin{aligned} \pi_2\left(q^m\right) &= \pi_2\left(\left(\pi_1\left(q^{m-1}\right) + \pi_2\left(q^{m-1}\right) j\right)(z_1 + z_2 j)\right) \\ &= z_2 \pi_1\left(q^{m-1}\right) + \pi_2\left(q^{m-1}\right) \bar{z}_1, \end{aligned} \tag{32}$$

we get the thesis by induction on m. □

Let $x \in \mathbb{H}_3 \setminus \mathbb{R}$ be fixed. From Theorem 3 we get that $(\mathrm{d}q^m)_x \in \mathrm{Hol}_p(\mathbb{H}, \mathbb{H})$, where $p := \mathrm{Im}(x)/|\mathrm{Im}(x)|$. Since $f \in \mathcal{SR}(\Omega_D)$ has a series expansion, we get that also $\mathrm{d}f_x$ belongs to $\mathrm{Hol}_p(\mathbb{H}, \mathbb{H})$ for every $x \in (\Omega_D \cap \mathbb{H}_3) \setminus \mathbb{R}$. To finish the proof of Theorem 1, we observe that $J_p = L_p$, since $x \in \mathbb{H}_3$. Therefore the linear map $\mathrm{d}f_x : (\mathbb{H}, J_p) \to (\mathbb{H}, J_p)$ is holomorphic. □

From the holomorphicity of the differentials $\mathrm{d}f_x$ at a reduced quaternion x, we get immediately the following properties for the real Jacobian matrix of f.

Corollary 4 *If $f \in \mathcal{SR}(\Omega_D)$ and $x \in \widetilde{\Omega}_D \setminus \mathbb{R} = (\Omega_D \cap \mathbb{H}_3) \setminus \mathbb{R}$, then*

(i) $\det(Jac(f)) \geq 0$ *at* x.
(ii) $\operatorname{rank}(Jac(f))$ *is even at* x.

The characterization of $\operatorname{Reg}(\mathcal{SR}(\Omega_D))$ in $\widetilde{\mathcal{R}}(\Omega_D)$ is completed by the following converse statement.

Proposition 5 *Assume Ω_D connected and $\Omega_D \cap \mathbb{R} \neq \emptyset$. Let $f \in \mathcal{R}(\Omega_D)$. If $\mathrm{d}f_x \in \operatorname{Hol}_p(\mathbb{H}, \mathbb{H})$ for every $x \in (\Omega_D \cap \mathbb{H}_3) \setminus \mathbb{R}$, with $p := \operatorname{Im}(x)/|\operatorname{Im}(x)|$ then there exists a (unique) slice regular function g on Ω_D, such that g and f are equal up to the first order on $\Omega_D \cap \mathbb{H}_3$.*

Proof For every $p \in \mathbb{S} \cap \mathbb{H}_3$ and any $x \in (\Omega_D \cap \mathbb{C}_p) \setminus \mathbb{R}$, the restriction $f_p = f_{|\mathbb{C}_p} : \mathbb{C}_p \to \mathbb{H}$ is holomorphic w.r.t. the structure J_p, since $\mathbb{C}_p$ is a complex subspace of $\mathbb{H}$ w.r.t. J_p and the differential $\mathrm{d}f_x \in \operatorname{Hol}_p(\mathbb{H}, \mathbb{H})$ by hypothesis. As proven in [10, Corollary 9], this implies that the restriction of f to $\Omega_D \cap \mathbb{H}_3$ is a *slice monogenic* function (cf. [4]) when $\mathbb{H}$ is identified with the Clifford algebra $\mathbb{R}_2$ and $\mathbb{H}_3$ is identified with the subspace of *paravectors* in $\mathbb{R}_2$. As seen in [10], every slice monogenic function on $\Omega_D \cap \mathbb{H}_3$ can be uniquely extended to a slice regular function on Ω_D. Since $\mathcal{D}(g - f) = 0$ on $\Omega_D \cap \mathbb{H}_3$, $g = f$ up to the first order on $\Omega_D \cap \mathbb{H}_3$. □

3 The Full Dirac Operators

We now look at the higher dimensional situation. Our starting point is the modified Cauchy-Riemann-Fueter operator $\mathcal{D}$. If we consider the quaternionic space as the real Clifford algebra $\mathbb{R}_2$, we can give a new look at $\mathcal{D}$ in terms of the algebraic involutions of the algebra. This reinterpretation of $\mathcal{D}$ suggests to study a new first order differential operator on the Clifford algebras $\mathbb{R}_n$, which behaves well w.r.t. monogenic functions and also w.r.t. the powers of the (complete) Clifford variable. This last property relates the operator with the theory of slice monogenic and slice regular functions.

3.1 The Operators $\mathcal{D}_n$

Denote by $e_1, \ldots, e_n$ the generators of $\mathbb{R}_n$. Let $x = \sum_K x_K e_K$ be a Clifford number, where $K = (i_1, \ldots, i_k)$ is a multiindex, with $0 \leq |K| := k \leq n$, the coefficients x_K are real numbers and e_K is the product of basis elements $e_K = e_{i_1} \cdots e_{i_k}$.

Definition 6 Let $\mathcal{D}_1 = \frac{1}{2}(\frac{\partial}{\partial x_0} + e_1 \frac{\partial}{\partial x_1})$ and $\mathcal{D}_{1,2} = \frac{1}{2}(\frac{\partial}{\partial x_2} + e_1 \frac{\partial}{\partial x_{12}})$. For $n > 1$, define recursively

$$\mathcal{D}_n := \mathcal{D}_{n-1} + e_n \mathcal{D}_{n-1,n}. \tag{33}$$

Here we consider $\mathbb{R}_{n-1}$ embedded in $\mathbb{R}_n$ and $\mathcal{D}_{n-1,n}$ is the operator defined as $\mathcal{D}_{n-1}$ w.r.t. the 2^{n-1} variables $x_n, x_{1n}, x_{2n}, \ldots, x_{12n}, \ldots, x_{12\cdots n}$. Since $\mathcal{D}_n$ depends on all the basis coordinates of $\mathbb{R}_n$, we call it the *full Dirac operator* on $\mathbb{R}_n$.

Remark 4 The operator $\mathcal{D}_1$ is the standard Cauchy-Riemann operator on the complex plane $\mathbb{C} \simeq \mathbb{R}_1$. The operator $\mathcal{D}_2$ is the same as the modified Cauchy-Riemann-Fueter operator $\mathcal{D}$ on $\mathbb{H} \simeq \mathbb{R}_2$:

$$\mathcal{D}_2 = \frac{1}{2}\left(\frac{\partial}{\partial x_0} + e_1 \frac{\partial}{\partial x_1} + e_2 \frac{\partial}{\partial x_2} - e_{12} \frac{\partial}{\partial x_{12}}\right). \tag{34}$$

$\mathcal{D}_3 = \mathcal{D}_2 + e_3 \mathcal{D}_{2,3}$ has the following expression

$$\mathcal{D}_3 = \frac{1}{2}\left(\frac{\partial}{\partial x_0} + e_1 \frac{\partial}{\partial x_1} + e_2 \frac{\partial}{\partial x_2} + e_3 \frac{\partial}{\partial x_3} - e_{12} \frac{\partial}{\partial x_{12}} - e_{13} \frac{\partial}{\partial x_{13}} \right.$$
$$\left. - e_{23} \frac{\partial}{\partial x_{23}} - e_{123} \frac{\partial}{\partial x_{123}}\right).$$

Despite its recursive definition, the operator $\mathcal{D}_n$ is symmetric w.r.t. the basis elements $e_1, \ldots, e_n$. More precisely, it has the following expression involving the *reversion* anti-involution of $\mathbb{R}_n$.

Proposition 6 *The operator $\mathcal{D}_n$ can be written in the following form*:

$$\mathcal{D}_n = \frac{1}{2} \sum_{|K| \le n} e_K^* \frac{\partial}{\partial x_K} \tag{35}$$

where $e_K^* = (-1)^{\frac{k(k-1)}{2}} e_K$ is obtained by applying to e_K the reversion anti-involution $x \mapsto x^*$. Moreover,

$$\mathcal{D}_{n-1,n} = \frac{1}{2} \sum_{H \not\ni n} e_H^* \frac{\partial}{\partial x_{(Hn)}}. \tag{36}$$

Proof $\mathcal{D}_1$ and $\mathcal{D}_{1,2}$ have the required form. Equations (35) and (36) follow from an easy inductive argument. □

On functions depending only on paravectors, the operator $\mathcal{D}_n$ acts as $\frac{1}{2}\mathcal{W}_n$, where $\mathcal{W}_n$ is the *Weyl* (or *Cauchy-Riemann*) *operator* on $\mathbb{R}_n$:

$$\mathcal{W}_n = \frac{\partial}{\partial x_0} + \sum_{i=1}^{n} e_i \frac{\partial}{\partial x_i}. \tag{37}$$

Corollary 5 *Every monogenic function (i.e. in the kernel of $\mathcal{W}_n$) defined on an open subset of the paravector subspace $\mathbb{R}^{n+1} \subset \mathbb{R}_n$ can be identified with an element of* $\ker \mathcal{D}_n$.

We can define also the conjugated operator $\overline{\mathcal{D}}_n$ and the auxiliary operator $\mathcal{D}_n^*$.

Definition 7

$$\begin{cases} \overline{\mathcal{D}}_1 = \dfrac{\partial}{\partial z_1} = \dfrac{1}{2}\left(\dfrac{\partial}{\partial x_0} - e_1 \dfrac{\partial}{\partial x_1}\right), & \mathcal{D}_1^* = \dfrac{\partial}{\partial \bar{z}_1} = \dfrac{1}{2}\left(\dfrac{\partial}{\partial x_0} + e_1 \dfrac{\partial}{\partial x_1}\right) \\ \overline{\mathcal{D}}_n = \overline{\mathcal{D}}_{n-1} - e_n \mathcal{D}_{n-1,n}^*, & \mathcal{D}_n^* = \mathcal{D}_{n-1}^* + e_n \overline{\mathcal{D}}_{n-1,n} \end{cases}$$

where $\mathcal{D}_{n-1,n}^*$ and $\overline{\mathcal{D}}_{n-1,n}$ are defined as $\mathcal{D}_{n-1}^*$ and $\overline{\mathcal{D}}_{n-1}$ w.r.t. the 2^{n-1} variables $x_n, x_{1n}, \ldots, x_{12n}, \ldots, x_{12\cdots n}$.

Still by induction, we obtain the following explicit forms for the operators $\overline{\mathcal{D}}_n$ and $\mathcal{D}_n^*$, now involving the *principal involution* of $\mathbb{R}_n$.

Proposition 7

$$\overline{\mathcal{D}}_n = \frac{1}{2} \sum_{|K| \le n} \tilde{e}_K \frac{\partial}{\partial x_K}, \tag{38}$$

where $\tilde{e}_K = (-1)^k e_K$ is obtained by applying to e_K the principal involution $x \mapsto \tilde{x}$. Moreover,

$$\mathcal{D}_n^* = \frac{1}{2} \sum_{|K| \le n} e_K \frac{\partial}{\partial x_K} \quad \text{and} \quad \mathcal{D}_{n-1,n}^* = \frac{1}{2} \sum_{H \not\ni n} e_H \frac{\partial}{\partial x_{Hn}}. \tag{39}$$

The differential operator

$$\mathcal{D}_n^* = \frac{1}{2} \sum_{|K| \le n} e_K \frac{\partial}{\partial x_K}$$

has already been considered in the literature (cf. [28] and [14]). The behavior of $\mathcal{D}_n$ w.r.t. power functions (see below Theorem 5) and the property stated in the next remark indicate that the operators $\mathcal{D}_n$ are better suited than $\mathcal{D}_n^*$ or $\overline{\mathcal{D}}_n$ to the theory of polynomials or more generally *slice regular* functions on a Clifford algebra. In [29] Dirac operators on the subspace of l-vectors have been studied. They coincide (up to sign) with the restriction of $\mathcal{D}_n$ to l-vectors. Since we are interested in the global behavior of the operator on the algebra, the choice of the grade-depending sign for the coefficients of $\mathcal{D}_n$ is essential.

Remark 5 The identity function x of $\mathbb{R}_3$ belongs to the kernel of $\mathcal{D}_2$ and $\mathcal{D}_3$ (and of course of the Cauchy-Riemann operator $\mathcal{D}_1$). Starting from $\mathcal{D}_1 x = 0$, $\mathcal{D}_{1,2} x = 0$ on $\mathbb{R}_2$, we get recursively that $\mathcal{D}_n x = 0$ on $\mathbb{R}_n$ for every n. Observe that even if $\mathcal{D}_1^* x = \mathcal{D}_{1,2}^* x = 0$, the identity function does not belong to the kernels of $\mathcal{D}_n^*$ or $\overline{\mathcal{D}}_n$ for every n.

3.2 Slice Regularity and the Full Dirac Operators

We are interested in the values of $\mathcal{D}_n$ on polynomials $\sum_m x^m a_m$ in the complete Clifford variable x. We start from the powers of x. To express our computation, we need some definitions and results from the theory of *slice regular* functions on $\mathbb{R}_n$ (see [9, 10] where the theory of slice regularity is constructed in a greater generality, for functions defined on a real alternative *-algebra).

Definition 8 Let $t(x) = x + \bar{x}$ be the *trace* of x and $n(x) = x\bar{x}$ the (squared) *norm* of $x \in \mathbb{R}_n$. The *quadratic cone* of $\mathbb{R}_n$ is the subset

$$\mathcal{Q}_n := \mathbb{R} \cup \{x \in \mathbb{R}_n \mid t(x) \in \mathbb{R},\ n(x) \in \mathbb{R},\ 4n(x) > t(x)^2\}.$$

(It can be seen that the last condition is automatically satisfied on $\mathbb{R}_n$.)

Let $\mathbb{S}_n := \{J \in \mathcal{Q}_n \mid J^2 = -1\} = \{x \in \mathbb{R}_n \mid t(x) = 0,\ n(x) = 1\}$. The elements of $\mathbb{S}_n$ are called *square roots of* -1 in the algebra $\mathbb{R}_n$.

Proposition 8 ([9, 10]) *The quadratic cone $\mathcal{Q}_n$ satisfies the following properties*:

(i) $\mathcal{Q}_n = \mathbb{R}_n$ *only for* $n = 1, 2$.

(ii) $\mathcal{Q}_n$ *contains (properly) the subspace of paravectors*

$$\mathbb{R}^{n+1} := \left\{x = \sum_K x_K e_K \in \mathbb{R}_n \,\middle|\, x_K = 0 \text{ for every } K \text{ such that } |K| > 1\right\}.$$

(iii) $\mathcal{Q}_n$ *is the real algebraic subset (proper for $n > 2$) of $\mathbb{R}_n$ defined by the equations*

$$x_K = 0, \qquad x \cdot (x e_K) = 0 \quad \forall e_K \neq 1 \text{ such that } e_K^2 = 1, \tag{40}$$

where $x \cdot y$ denotes the Euclidean scalar product on $\mathbb{R}_n \simeq \mathbb{R}^{2^n}$.

(iv) *For $J \in \mathbb{S}_n$, let $\mathbb{C}_J = \langle 1, J\rangle \simeq \mathbb{C}$ be the subalgebra generated by J. Then*

$$\mathcal{Q}_n = \bigcup_{J \in \mathbb{S}_n} \mathbb{C}_J \tag{41}$$

and $\mathbb{C}_I \cap \mathbb{C}_J = \mathbb{R}$ for every $I, J \in \mathbb{S}_n$, $I \neq \pm J$. As a consequence, if x belongs to $\mathcal{Q}_n$, also the powers x^m belong to the quadratic cone $\mathcal{Q}_n$.

Slice regular functions are defined only on subdomains of the quadratic cone (we refer to [9, 10] for full details). However, if the domain intersects the real axis, then the class of slice regular functions coincides with the one of functions having local power series expansion centered at real points.

Now we compute the values of $\mathcal{D}_n(x^m)$. We already know the result for $n = 2$ (cf. Corollary 1): $\mathcal{D}_2(x^m) = 0$ on the subset of reduced quaternions $\mathbb{H}_3 \subset \mathbb{H} \simeq \mathbb{R}_2$. We can show that the behavior of $\mathcal{D}_n$ on the powers depends on the parity of n (as many other properties of $\mathbb{R}_n$). In this scheme the quaternions ($n = 2$) are in some sense exceptional.

Theorem 5 *Let $x = \sum_{|K| \le n} x_K e_K$ be the complete Clifford variable in $\mathbb{R}_n$. The following facts hold*:

(i) *If n is an odd integer, then $\mathcal{D}_n(x^m) = 0$ on the whole algebra $\mathbb{R}_n$ for every integer $m \geq 1$.*
(ii) *If n is an even integer, $n > 2$, then $\mathcal{D}_n(x^m) = 0$ on the quadratic cone $\mathcal{Q}_n$ of $\mathbb{R}_n$ for every integer $m \geq 1$.*
(iii) *$\mathcal{D}_2(x^m) = 0$ on the subset of reduced quaternions $\mathbb{H}_3 \subset \mathbb{R}_2$ for every integer $m \geq 1$.*

In the proof of Theorem 5 we will apply the following algebraic lemma:

Lemma 2 *Let $N = (1, \ldots, n)$. For every $x \in \mathbb{R}_n$, it holds*

$$\sum_{H \not\ni n} e_H^* x e_H = 2^{n-1}(x_n e_n + x_{(1\cdots n-1)} e_{(1\cdots n-1)}) \quad \textit{for odd } n, \tag{42}$$

$$\sum_{H \not\ni n} e_H^* x e_H = 2^{n-1}(x_n e_n + x_N e_N) \quad \textit{for even } n. \tag{43}$$

Proof Let $h = |H|$, $k = |K|$ and let $\sigma_{H,K}$ be the sign such that $e_H e_K = \sigma_{H,K} e_K e_H$. Then it holds

$$\begin{aligned} e_H^* x e_H &= (-1)^{\frac{h(h-1)}{2}} \sum_K x_K \sigma_{H,K} e_K e_H^2 = (-1)^{\frac{h(h-1)}{2}} \sum_K x_K \sigma_{H,K} e_K (-1)^{\frac{h(h+1)}{2}} \\ &= (-1)^h \sum_K x_K \sigma_{H,K} e_K. \end{aligned} \tag{44}$$

If i is the cardinality of $H \cap K$, then $\sigma_{H,K} = (-1)^{hk+i}$. Therefore

$$\sum_{H \not\ni n} e_H^* x e_H = \sum_K \left(\sum_{H \not\ni n} (-1)^{h(k+1)+i} \right) x_K e_K. \tag{45}$$

If k is even, then $\sum_{H \not\ni n} (-1)^{h(k+1)+i} = \sum_{H \not\ni n} (-1)^{h+i} = \sum_{H \not\ni n} (-1)^{h-i}$ counts the difference between the number of the even and the odd subsets of the set $\{1, \ldots, n-1\} \setminus K$. Therefore the sum is zero unless n is odd and $K = (1, \ldots, n-1)$ or n is even and $K = (1, \ldots, n)$. In both cases the sum is equal to 2^{n-1}.

If k is odd, then $\sum_{H \not\ni n} (-1)^{h(k+1)+i} = \sum_{H \not\ni n} (-1)^i$ counts the difference between the number of even and odd subsets of $K \cap \{1, \ldots, n-1\}$. Then it is zero unless $K = (n)$. In this case, the sum is 2^{n-1}.

From these and (45) we get the statement of the lemma. □

Proof of Theorem 5 The third case ($n = 2$) has already been proved in Corollary 1.

Case (i): n *odd*. We show by induction on m that $\mathcal{D}_{n-1} x^m = -e_n \mathcal{D}_{n-1,n} x^m$. Since $\mathcal{D}_n x = 0$ (cf. Remark 5), the equality is valid for $m = 1$. Take $m > 1$ and assume that $\mathcal{D}_{n-1} x^{m-1} = -e_n \mathcal{D}_{n-1,n} x^{m-1}$. We have the following product formula (obtained in a way similar to the $n = 2$ case of Proposition 3):

$$\begin{aligned} \mathcal{D}_{n-1,n} x^m &= \left(\mathcal{D}_{n-1,n} x^{m-1}\right) x + \frac{1}{2} \sum_{H \not\ni n} e_H^* x^{m-1} e_{(Hn)} \\ &= \left(\mathcal{D}_{n-1,n} x^{m-1}\right) x + \frac{1}{2} \left(\sum_{H \not\ni n} e_H^* x^{m-1} e_H \right) e_n. \end{aligned} \tag{46}$$

Since, from Lemma 2,

$$\Big(\sum_{H\not\ni n} e_H^* x e_H\Big) e_n = 2^{n-1}(-x_n + x_{(1\cdots n-1)} e_N), \tag{47}$$

the last term in Eq. (46) belongs to the center $\langle 1, e_N\rangle$ of $\mathbb{R}_n$. Therefore, from (46) we get

$$-e_n \mathcal{D}_{n-1,n} x^m = -e_n\big(\mathcal{D}_{n-1,n} x^{m-1}\big)x + \frac{1}{2}\sum_{H\not\ni n} e_H^* x^{m-1} e_H. \tag{48}$$

On the other hand, we also have

$$\mathcal{D}_{n-1} x^m = \big(\mathcal{D}_{n-1} x^{m-1}\big)x + \frac{1}{2}\sum_{H\not\ni n} e_H^* x^{m-1} e_H \tag{49}$$

and then the inductive hypothesis gives the equality $\mathcal{D}_{n-1}x^m = -e_n\mathcal{D}_{n-1,n}x^m$, which is equivalent to $\mathcal{D}_n x^m = 0$.

Case (ii): *n even*, greater than 2. We show that

$$\mathcal{D}_n x^m = \big(\mathcal{D}_n x^{m-1}\big)x + 2^{n-1}\big[x^{m-1}\big]_N e_N, \tag{50}$$

where $[a]_N$ denotes the coefficient of the *pseudoscalar* e_N of the element $a \in \mathbb{R}_n$. If $m = 1$, the equality (50) is true since $\mathcal{D}_n x = 0$. Let $m > 1$. Then it holds

$$\begin{aligned}\mathcal{D}_n x^m &= \mathcal{D}_{n-1} x^m + e_n \mathcal{D}_{n-1,n} x^m\\ &= \big(\mathcal{D}_{n-1} x^{m-1}\big)x + \frac{1}{2}\sum_{H\not\ni n} e_H^* x^{m-1} e_H\\ &\quad + e_n\Big(\big(\mathcal{D}_{n-1,n} x^{m-1}\big)x + \frac{1}{2}\sum_{H\not\ni n} e_H^* x^{m-1} e_H e_n\Big).\end{aligned} \tag{51}$$

From Lemma 2, since n is even we have

$$\sum_{H\not\ni n} e_H^* x^{m-1} e_H + e_n \sum_{H\not\ni n} e_H^* x^{m-1} e_H e_n = 2^n\big[x^{m-1}\big]_N e_N \tag{52}$$

and therefore, from (51) and (52)

$$\mathcal{D}_n x^m = \big(\mathcal{D}_n x^{m-1}\big)x + 2^{n-1}\big[x^{m-1}\big]_N e_N. \tag{53}$$

Now we prove by induction on m that $\mathcal{D}_n x^m$ vanishes on the quadratic cone $\mathcal{Q}_n$. For $m = 1$, $\mathcal{D}_n x = 0$ on the whole algebra. Let $m > 1$ and assume that $\mathcal{D}_n x^{m-1} = 0$ at every point of $\mathcal{Q}_n$. Since the power function maps $\mathcal{Q}_n$ in $\mathcal{Q}_n$, for every $x \in \mathcal{Q}_n$ we have $[x^{m-1}]_N = 0$. The equality (53) and the inductive hypothesis allow to conclude that $\mathcal{D}_n x^m = 0$ at $x \in \mathcal{Q}_n$. □

From the right linearity of the operators $\mathcal{D}_n$, we get the following result.

Corollary 6 *Let $n \geq 3$. Let $p(x) = \sum_{m=0}^{d} x^m a_m$ be a polynomial in the complete Clifford variable $x = \sum_{|K| \leq n} x_K e_K$ with right Clifford coefficients. If n is odd, then p is in the kernel of $\mathcal{D}_n$. If n is even, then the restriction of $\mathcal{D}_n(p)$ to the quadratic cone $\mathcal{Q}_n$ vanishes.*

Polynomials $p(x) = \sum_{m=0}^{d} x^m a_m$ and convergent power series $\sum_k x^k a_k$ with right Clifford coefficients are examples of *slice regular* functions on the intersection of $\mathcal{Q}_n$ with a ball centered in the origin (cf. [9, 10]). If $n \geq 3$, slice regularity generalizes the concept of *slice monogenic functions* introduced in [4]: if f is slice regular on a domain which intersects the real axis, then the restriction of f to the set of paravectors is a slice monogenic function. Conversely, every slice monogenic function is the restriction of a unique slice regular function. Since every slice monogenic function has a power expansions in the paravector variable, centered at points of the real axis (cf. [4]), every slice monogenic f function has an extension $\tilde{f}$ to on open domain in $\mathbb{R}_n$ which satisfies the property stated in Corollary 6: if n is odd, then $\tilde{f}$ is in the kernel of $\mathcal{D}_n$. If n is even, then the restriction of $\mathcal{D}_n(\tilde{f})$ to the quadratic cone $\mathcal{Q}_n$ vanishes.

The same property holds for slice regular functions on a domain Ω in $\mathcal{Q}_n$ with non empty intersection with $\mathbb{R}$. If f is slice regular then it has local power series expansion (in the complete Clifford variable) on an open neighborhood of every real point. This can be seen using the *Clifford operator norm* (see [11, 7.20]) of $\mathbb{R}_n$, which reduces to the Clifford norm on the quadratic cone $\mathcal{Q}_n$.

Remark 6 For $n = 1, 2$ the operators $\mathcal{D}_n$ are elliptic, since in this case

$$4\overline{\mathcal{D}}_n \mathcal{D}_n = 4\mathcal{D}_n \overline{\mathcal{D}}_n = \Delta_{\mathbb{R}^{2^n}}. \tag{54}$$

For $n = 3$ it holds

$$4\overline{\mathcal{D}}_3 \mathcal{D}_3 = 4\mathcal{D}_3 \overline{\mathcal{D}}_3 = \Delta_{\mathbb{R}^8} + \mathcal{L}_3, \tag{55}$$

where

$$\mathcal{L}_3 = -2\left(\frac{\partial}{\partial x_0}\frac{\partial}{\partial x_{123}} - \frac{\partial}{\partial x_1}\frac{\partial}{\partial x_{23}} + \frac{\partial}{\partial x_2}\frac{\partial}{\partial x_{13}} - \frac{\partial}{\partial x_3}\frac{\partial}{\partial x_{13}}\right)e_{123}. \tag{56}$$

For $n \geq 4$,

$$4\overline{\mathcal{D}}_n \mathcal{D}_n = \Delta_{\mathbb{R}^{2^n}} + \mathcal{L}_n \quad \text{and} \quad 4\mathcal{D}_n \overline{\mathcal{D}}_n = \Delta_{\mathbb{R}^{2^n}} + \mathcal{L}'_n, \tag{57}$$

where

$$\mathcal{L}_n = \sum_{H \neq K} t(e_H^* \tilde{e}_K)\frac{\partial}{\partial x_H}\frac{\partial}{\partial x_K} \quad \text{and} \quad \mathcal{L}'_n = \sum_{H \neq K} t(\tilde{e}_H e_K^*)\frac{\partial}{\partial x_H}\frac{\partial}{\partial x_K} \tag{58}$$

(the summations are made over multiindices H, K without repetitions). For $n \geq 4$ the operators $\mathcal{L}_n$ and $\mathcal{L}'_n$ are different. In particular, for $n \geq 3$ the operators $\mathcal{D}_n$ are not elliptic. Note that the symbol of the differential operator $\mathcal{L}_3$ is, up to a multiplicative constant, the polynomial $x_0 x_{123} - x_1 x_{23} + x_2 x_{13} - x_3 x_{12}$ whose zero set is the *normal cone* of the Clifford algebra $\mathbb{R}_3$ (cf. [10] for its definition). A similar relation holds for the symbols of $\mathcal{L}_n$ and $\mathcal{L}'_n$ and the equations of the normal cone of $\mathbb{R}_n$ for $n > 3$.

3.3 The Case of $\mathfrak{D}_3$

In $\mathbb{R}_3$ can be introduced a particular algebraic decomposition in terms of paravector variables. Denote by $I = e_{123}$ the pseudoscalar of $\mathbb{R}_3$. The central idempotents $I_\pm = \frac{1}{2}(1 \pm I)$ satisfy the properties

$$I_+^2 = I_+, \qquad I_-^2 = I_-, \qquad I_+I_- = I_-I_+ = 0, \qquad I_+ + I_- = 1. \tag{59}$$

Let $X = x_0 + x_1e_1 + x_2e_2 + x_3e_3$ be the paravector variable and $X' = x - X = x_{12}e_{12} + x_{13}e_{13} + x_{23}e_{23} + x_{123}e_{123}$. We can define two new (rotated) paravector variables $Y = y_0 + y_1e_1 + y_2e_2 + y_3e_3$ and $Z = z_0 + z_1e_1 + z_2e_2 + z_3e_3$ by setting

$$Y = \frac{1}{2}\big(X + X'I\big), \qquad Z = \frac{1}{2}\big(X - X'I\big), \tag{60}$$

from which we get the decomposition

$$x = X + X' = Y + Z + (Y - Z)I = 2YI_+ + 2ZI_-. \tag{61}$$

Since the multiplication by $I_\pm$ gives two orthogonal projections, for every positive integer m it holds

$$x^m = (2Y)^m I_+ + (2Z)^m I_-, \tag{62}$$

and therefore for every polynomial, power series or in general for a slice regular function f on a domain which intersects the real axis, we can write

$$f(x) = f(2Y)I_+ + f(2Z)I_-. \tag{63}$$

The operator $\mathfrak{D}_3$ decomposes as $\mathfrak{D}_3 = \frac{1}{2}(\partial_X - \partial_{X'})$, where $\partial_X = \partial_{x_0} + e_1\partial_{x_1} + e_2\partial_{x_2} + e_3\partial_{x_3}$ is the Weyl operator of $\mathbb{R}_3$ and $\partial_{X'} = e_{12}\partial_{x_{12}} + e_{13}\partial_{x_{13}} + e_{23}\partial_{x_{23}} + e_{123}\partial_{x_{123}}$. Denote by ∂_Y and ∂_Z the Weyl operators w.r.t. Y and Z respectively. Then

$$\partial_X = \frac{1}{2}(\partial_Y + \partial_Z), \qquad \partial_{X'} = \frac{1}{2}(\partial_Y - \partial_Z)I, \tag{64}$$

and therefore in the variables Y, Z the operator $\mathfrak{D}_3$ has the following form:

$$\mathfrak{D}_3 = I_-\partial_Y + I_+\partial_Z = \partial_Y I_- + \partial_Z I_+. \tag{65}$$

This decomposition implies that a function f belongs to the kernel of $\mathfrak{D}_3$ if and only if its projections $f_- := fI_-$ and $f_+ := fI_+$ belong to the kernels of the Weyl operators ∂_Y and ∂_Z respectively. In particular, every pair of arbitrary functions $g(Y)$, $h(Z)$ define a function $f(Y, Z) = I_-h(Z) + I_+g(Y)$ in the kernel of $\mathfrak{D}_3$. This property shows again that $\mathfrak{D}_3$ is not an elliptic operator, as can be seen also when formula (55) is expressed in the variables Y, Z:

$$4\overline{\mathfrak{D}}_3\mathfrak{D}_3 = 4\mathfrak{D}_3\overline{\mathfrak{D}}_3 = \frac{1}{2}(\Delta_Y + \Delta_Z) - \frac{1}{2}(\Delta_Y - \Delta_Z)I = I_-\Delta_Y + I_+\Delta_Z, \tag{66}$$

where Δ_Y is the Laplacian w.r.t. the variables y_0, y_1, y_2, y_3 and similarly for Δ_Z.

3.4 The Space $\mathcal{F}(\Omega)$

In view of the non-ellipticity of $\mathcal{D}_3$, we consider a proper subspace of $\ker \mathcal{D}_3$. As we will see in Corollary 7, this space extends the one of monogenic functions. Consider the Laplacians $\Delta_X = \partial_X \overline{\partial}_X$, $\Delta_{X'} = \partial_{X'} \overline{\partial}_{X'}$ and $\Delta = \partial_X \overline{\partial}_X + \partial_{X'} \overline{\partial}_{X'} = \Delta_{\mathbb{R}^8}$.

Definition 9 Let Ω be an open subset of $\mathbb{R}_3$. We define

$$\mathcal{F}(\Omega) := \left\{ f \in C^1(\Omega) \mid \mathcal{D}_3 f = 0,\ \Delta_X \partial_X f = 0 \text{ on } \Omega \right\}.$$

The space $\mathcal{F}(\Omega)$ can be expressed in the paravector variables Y, Z in the way described by the next proposition.

Proposition 9 *Let $\Omega \subseteq \mathbb{R}_3$ be open. Then*

$$\mathcal{F}(\Omega) = \left\{ f \in C^1(\Omega) \mid \mathcal{D}_3 f = 0,\ \Delta_Y \partial_Y f = \Delta_Z \partial_Z f = 0 \text{ on } \Omega \right\}.$$

Every $f \in \mathcal{F}(\Omega)$ is biharmonic on Ω (i.e. $\Delta^2 f = 0$) and also biharmonic w.r.t. the variables Y and Z separately. In particular, it is real analytic on Ω. Moreover, $f = f_- + f_+ \in \mathcal{F}(\Omega)$ if and only if its projections f_- and f_+ satisfy

$$\partial_Y f_- = \Delta_Z \partial_Z f_- = 0, \qquad \partial_Z f_+ = \Delta_Y \partial_Y f_+ = 0. \tag{67}$$

Proof If $\mathcal{D}_3 f = 0$, then $\partial_X f = \partial_{X'} f$. Therefore $\Delta f = (\partial_X \overline{\partial}_X + \partial_{X'} \overline{\partial}_{X'}) f = 2\Delta_X f$. Moreover, from (64) it follows that $\partial_Z f = (\partial_X - I\partial_{X'}) f = (\partial_X - I\partial_X) f = 2I_- \partial_X f$. Then

$$\Delta_Z f = \partial_Z \overline{\partial}_Z f = 4I_- \partial_X \overline{\partial}_X f = 4\Delta_X f_- = 2\Delta f_- \tag{68}$$

and therefore $\Delta_Z \partial_Z f = 8\Delta_X \partial_X f_-$. A similar computation gives $\Delta_Y \partial_Y f = 8\Delta_X \partial_X f_+$. Then $\Delta_X \partial_X f = 0$ if and only if $\Delta_Z \partial_Z f = \Delta_Y \partial_Y f = 0$.

If $f \in \mathcal{F}(\Omega)$, then $0 = \overline{\partial}_Z \partial_Z \Delta_Z f = \Delta_Z^2 f$ and $0 = \overline{\partial}_Y \partial_Y \Delta_Y f = \Delta_Y^2 f$. From these equalities we get $4\Delta^2 f_- = \Delta_Z^2 f = 0$, $4\Delta^2 f_+ = \Delta_Y^2 f = 0$ and then $\Delta^2 f = 0$: f is biharmonic on Ω.

The last statement is immediate from the decomposition (65) of $\mathcal{D}_3$. □

Remark 7 The preceding proposition tells that every function in the space $\mathcal{F}(\Omega)$ is (separately) *holomorphic Cliffordian* [17] in the paravector variables X, Y and Z.

Corollary 7 *Every polynomial $p(x) = \sum_{m=0}^{d} x^m a_m$ in the complete Clifford variable $x = \sum_{|K| \le 3} x_K e_K$ belongs to $\mathcal{F}(\mathbb{R}_3)$. The same holds for every slice regular function on a domain in the quadratic cone $\mathcal{Q}_3$ intersecting the real axis. If $f(X)$ is a function depending only on the paravector variable X of $\mathbb{R}_3$, then $f \in \mathcal{F}$ if and only if it is monogenic, i.e. $\partial_X f = 0$.*

Proof From the algebraic decomposition (62), every power of x can be expressed by means of powers of Y and Z. Since every power of a paravector variable X is holomorphic Cliffordian (cf. [17]), i.e. $\Delta_X \partial_X f = 0$, the first two statements follow from Theorem 5 and Corollary 6. The last statement is an immediate consequence of Corollary 5. □

Let B denote the eight-dimensional unit ball in $\mathbb{R}_3$. Let $T \simeq S^3 \times S^3$ be the subset of the unit sphere ∂B defined by $T := \{|Y| = |Z| = 1/2\}$ and $P := \{|Y| < 1/2\} \cap \{|Z| < 1/2\}$. Since $|x|^2 = 2|Y|^2 + 2|Z|^2$, $P \subset B$ and $T \subset \partial P$. We will call T the *distinguished boundary* of P. Note that T is contained in the normal cone $\mathcal{N}_3$ of $\mathbb{R}_3$ (cf. [10]), which has equation $|Y| = |Z|$ in the variables Y, Z.

Proposition 10 (Integral Representation Formula) *There is an integral representation formula for functions $f \in \mathcal{F}(P) \cap C^2(\overline{P})$ with the distinguished boundary T as domain of integration. The values of f on P are determined by the values on T of f, $\partial_X f$ and the second derivatives $\frac{\partial}{\partial x_K}(\partial_X f)$ for multiindices K with $|K| \le 3$.*

Proof Consider the component $f_- \in \mathcal{F}(P)$. Since $\partial_Y f_- = 0$, we can apply the representation formula for the Weyl operator ∂_Y (cf. [1]) and reconstruct f_- on the set $\{|Y| < 1/2, |Z| = 1/2\}$. On functions in the class $\mathcal{F}(P)$, the operators ∂_Y and ∂_Z commute (see the proof of Proposition 9). Since $\partial_Y \partial_Z f_- = \partial_Z \partial_Y f_- = 0$, we can reconstruct also $\partial_Z f_-$ on the set $\{|Y| < 1/2, |Z| = 1/2\}$. Since $\Delta_Z \partial_Z f_- = 0$, we can now apply the integral representation formula for holomorphic Cliffordian functions (see [17]) w.r.t. the paravector variable Z and obtain the values of f_- on P. A similar reasoning for f_+ gives the result. □

Acknowledgements This work was partially supported by MIUR (PRIN Project "Proprietà geometriche delle varietà reali e complesse") and GNSAGA of INdAM.

References

1. Brackx, F., Delanghe, R., Sommen, F.: Clifford Analysis. Research Notes in Mathematics, vol. 76. Pitman (Advanced Publishing Program), Boston (1982)
2. Chen, J., Li, J.: Quaternionic maps between hyperkähler manifolds. J. Differ. Geom. **55**(2), 355–384 (2000)
3. Colombo, F., Sabadini, I., Sommen, F., Struppa, D.C.: Analysis of Dirac Systems and Computational Algebra. Progress in Mathematical Physics, vol. 39. Birkhäuser Boston, Boston (2004)
4. Colombo, F., Sabadini, I., Struppa, D.C.: Slice monogenic functions. Isr. J. Math. **171**, 385–403 (2009)
5. Folland, G.B.: Introduction to Partial Differential Equations, 2nd edn. Princeton University Press, Princeton (1995)
6. Fueter, R.: Die Funktionentheorie der Differentialgleichungen $\Delta u = 0$ und $\Delta\Delta u = 0$ mit vier reellen Variablen. Comment. Math. Helv. **7**(1), 307–330 (1934)
7. Gentili, G., Struppa, D.C.: A new approach to Cullen-regular functions of a quaternionic variable. C. R. Math. Acad. Sci. Paris **342**(10), 741–744 (2006)
8. Gentili, G., Struppa, D.C.: A new theory of regular functions of a quaternionic variable. Adv. Math. **216**(1), 279–301 (2007)

9. Ghiloni, R., Perotti, A.: A new approach to slice regularity on real algebras. In: Hypercomplex Analysis and Its Applications. Trends in Mathematics. Birkhäuser, Basel (2010)
10. Ghiloni, R., Perotti, A.: Slice regular functions on real alternative algebras. Adv. Math. **226**(2), 1662–1691 (2011)
11. Gilbert, J.E., Murray, M.A.M.: Clifford Algebras and Dirac Operators in Harmonic Analysis. Cambridge Studies in Advanced Mathematics, vol. 26. Cambridge University Press, Cambridge (1991). doi:10.1017/CBO9780511611582
12. Gürlebeck, K., Habetha, K., Sprößig, W.: Holomorphic Functions in the Plane and n-Dimensional Space. Birkhäuser, Basel (2008)
13. Hengartner, W., Leutwiler, H.: Hyperholomorphic functions in $\mathbb{R}^3$. Adv. Appl. Clifford Algebras **11**(S1), 247–259 (2001)
14. Hestenes, D., Sobczyk, G.: Clifford Algebra to Geometric Calculus: A Unified Language for Mathematics and Physics. Fundamental Theories of Physics. Reidel, Dordrecht (1984)
15. Joyce, D.: Hypercomplex algebraic geometry. Quart. J. Math. Oxford Ser. (2) **49**(194), 129–162 (1998). doi:10.1093/qjmath/49.194.129
16. Kravchenko, V.V., Shapiro, M.V.: Integral Representations for Spatial Models of Mathematical Physics. Pitman Research Notes in Mathematics Series, vol. 351. Longman, Harlow (1996)
17. Laville, G., Ramadanoff, I.: Holomorphic Cliffordian functions. Adv. Appl. Clifford Algebras **8**(2), 323–340 (1998). doi:10.1007/BF03043103
18. Lichnerowicz, A.: Applications harmoniques et variétés kähleriennes. In: Symposia Mathematica, vol. III, INDAM, Rome, 1968/1969, pp. 341–402. Academic Press, London (1968/1969)
19. Naser, M.: Hyperholomorphe funktionen. Sib. Math. J. **12**, 959–968 (1971)
20. Nōno, K.: Characterization of domains of holomorphy by the existence of hyper-conjugate harmonic functions. Rev. Roum. Math. Pures Appl. **31**(2), 159–161 (1986)
21. Perotti, A.: Quaternionic regularity and the $\overline{\partial}$-Neumann problem in $\mathbb{C}^2$. Complex Var. Elliptic Equ. **52**(5), 439–453 (2007). doi:10.1080/17476930601178392
22. Perotti, A.: Every biregular function is a biholomorphic map. Adv. Appl. Clifford Algebras **19**(2), 441–451 (2009). doi:10.1007/s00006-009-0161-8
23. Perotti, A.: Holomorphic functions and regular quaternionic functions on the hyperkähler space $\mathbb{H}$. In: More Progresses in Analysis: Proceedings of the 5th International Isaac Congress. Trends in Mathematics, pp. 1009–1018. World Scientific, Singapore (2009)
24. Perotti, A.: Regular quaternionic functions and conformal mappings. CUBO **11**(1), 123–143 (2009)
25. Perotti, A.: On directional Hilbert operators for regular quaternionic functions on $\mathbb{R}^3$. Adv. Appl. Clifford Algebras **20**(3/4), 803–817 (2010)
26. Perotti, A.: The full Dirac operator on a Clifford algebra. In: Gürlebeck, K. (ed.) Proceedings of the 9th International Conference on Clifford Algebras and Their Applications (ICCA9), Weimar, Germany (2011)
27. Qian, T.: Generalization of Fueter's result to $\mathbf{R}^{n+1}$. Atti Accad. Naz. Lincei, Cl. Sci. Fis. Mat. Nat., Rend. Lincei (9) Mat. Appl. **8**(2), 111–117 (1997)
28. Rizza, G.B.: Funzioni regolari nelle algebre di Clifford. Rend. Mat. Appl. (5) **15**, 53–79 (1956)
29. Sabadini, I., Sommen, F.: Clifford analysis on the space of vectors, bivectors and l-vectors. In: Advances in Analysis and Geometry. Trends in Mathematics, pp. 161–185. Birkhäuser, Basel (2004)
30. Sce, M.: Osservazioni sulle serie di potenze nei moduli quadratici. Atti Accad. Naz. Lincei, Rend. Cl. Sci. Fis. Mat. Nat. (8) **23**, 220–225 (1957)
31. Shapiro, M.V., Vasilevski, N.L.: Quaternionic ψ-hyperholomorphic functions, singular integral operators and boundary value problems. I. ψ-hyperholomorphic function theory. Complex Var. Theory Appl. **27**(1), 17–46 (1995)
32. Sommen, F.: On a generalization of Fueter's theorem. Z. Anal. Anwend. **19**(4), 899–902 (2000)
33. Sudbery, A.: Quaternionic analysis. Math. Proc. Camb. Philos. Soc. **85**(2), 199–224 (1979)

A Note on Analytic Functionals on the Complex Light Cone

Daniele C. Struppa

Abstract I use Ehrenpreis' Fundamental Principle to reinterpret some results of Morimoto and Fujita (Hiroshima Math. J. 25:493–512, 1995) on analytic functionals on the complex light cone. I then show how these ideas can be used to generalize such results to the bicomplex and multicomplex setting.

1 Introduction and Notations

In this brief note, I would like to point out how the Fundamental Principle of Ehrenpreis [6] can be used to interpret, revisit, and generalize to the bicomplex and multicomplex setting some results obtained by Morimoto and Fujita in [8]. In this paper I will only sketch the general ideas and the possible directions to be pursued, and a full treatment of this topic will be given in the forthcoming [13].

Let us set a few notations. Let $z = (z_1, \dots, z_{n+1})$ be the variable in $\mathbb{C}^{n+1}$, and let $\Gamma := \{z \in \mathbb{C}^{n+1} : z_1^2 + \cdots + z_{n+1}^2 = 0\}$ be the so-called complex light cone. We will denote by $\mathscr{O}$ the sheaf of germs of holomorphic functions, so that $\mathscr{O}(\mathbb{C}^{n+1})$ is the space of entire functions on $\mathbb{C}^{n+1}$, and $\mathscr{O}(\Gamma)$ is the space of holomorphic functions on Γ. Note that this second space is defined as the quotient

$$\mathscr{O}(\Gamma) := \frac{\mathscr{O}(\mathbb{C}^{n+1})}{I}$$

where I is the ideal generated in $\mathscr{O}(\mathbb{C}^{n+1})$ by the polynomial $\Delta(z) := z_1^2 + \cdots + z_{n+1}^2$. The reason for the use of the letter Δ for this polynomial consists in the fact that such polynomial is in fact the symbol of the complex Laplacian, namely of the differential operator

$$\Delta := \frac{\partial^2}{\partial z_1^2} + \cdots + \frac{\partial^2}{\partial z_{n+1}^2},$$

D.C. Struppa (✉)
Schmid College of Science and Technology, Chapman University, One University Drive, Orange, CA 92866, USA
e-mail: struppa@chapman.edu

G. Gentili et al. (eds.), *Advances in Hypercomplex Analysis*, Springer INdAM Series 1, DOI 10.1007/978-88-470-2445-8_7,

acting on holomorphic functions. Another important space that will be needed is the space of entire functions of exponential type, and its analog on the variety Γ. In this case one has that the space of entire functions of exponential type is given by

$$\mathrm{Exp}(\mathbb{C}^{n+1}) := \{ f \in \mathcal{O}(\mathbb{C}^{n+1}) : |f(z)| \leq A \exp(B|z|), \text{ for some } A, B > 0 \},$$

with a similar definition for $\mathrm{Exp}(\Gamma)$. Finally, we recall that these spaces are all topological vector spaces, whose topologies are well known and are described for example in [14]; we can therefore study their strong duals (as customary, if X is a topological vector space, we will indicate by X' its strong dual), which are spaces of objects that can be considered as generalizations of analytic functionals. Specifically, $\mathcal{O}'(\mathbb{C}^{n+1})$ is the usual space of analytic functionals, while the dual $\mathrm{Exp}'(\mathbb{C}^{n+1})$ of $\mathrm{Exp}(\mathbb{C}^{n+1})$ is a space which properly contains analytic functionals. A similar commentary can be made for the dual of the spaces of holomorphic functions on the variety Γ.

In [8], the authors use a Fourier-Borel transform for such (generalized) functionals to prove some topological dualities. Specifically they show that

$$\mathcal{O}'(\Gamma) \cong \{ f \in \mathrm{Exp}(\mathbb{C}^{n+1}) : \Delta(f) = 0 \}, \tag{1}$$

and that

$$\mathrm{Exp}'(\Gamma) \cong \{ f \in \mathcal{O}(\mathbb{C}^{n+1}) : \Delta(f) = 0 \}. \tag{2}$$

In this paper I will show how these results can be seen in the framework of Ehrenpreis' Fundamental Principle, and I will outline a few related issues and connections with the theory of bicomplex (and multicomplex) numbers, that are ripe for discussion, and which will be addressed in more detail in [13].

2 Ehrenpreis' Fundamental Principle and the Duality Theorems

In 1960 Leon Ehrenpreis announced [5] his Fundamental Principle for Systems of Linear Constant Coefficients Partial Differential Operators (or Fundamental Principle in short), whose full proof was later given in [6] and independently in [9]. The Fundamental Principle applies to systems of such differential equations in what Ehrenpreis called Localizable Analytically Uniform Spaces (LAU-spaces). The technical definition of such spaces is rather complex, but for our purposes it suffices to think of them as topological vector spaces of generalized functions X such that their dual X' is topologically isomorphic (via some variation of the Fourier transform) to a space of entire functions satisfying suitable growth conditions at infinity. The theory of such spaces is described in detail in [6, 9], and a comprehensive survey is given in [2]. Among the many examples of LAU-spaces, two are important for our discussion. The first is the space $\mathcal{O}(\mathbb{C}^{n+1})$ of entire functions, and the second is the space $\mathrm{Exp}(\mathbb{C}^{n+1})$ of entire functions of exponential type. These two spaces are particularly important, and crucially connected, because they are the dual of each other, so that the space of analytic functionals is topologically isomorphic (via the Fourier-Borel transform) to the space of entire functions of exponential type. Similarly, the space of linear continuous functionals on the space of entire functions of exponential type is topologically isomorphic to the space of entire functions.

The importance of LAU-spaces was made evident by the Fundamental Principle, which can be stated as follows. Let X be an LAU-space of generalized functions in n variables, and let $P_1(D), \dots, P_r(D)$ be r differential operators with constant coefficients, whose symbols are the polynomials $P_1, \dots, P_r$. Let now $X^{\mathbf{P}}$ denote the space of generalized functions in X which are solutions of the system $P_1(D)f = \dots = P_r(D)f = 0$. The Fundamental Principle states that every element f in $X^{\mathbf{P}}$ can be represented as

$$f(x) = \sum_{k=1}^{t} \int_{V_k} Q_k(x) \exp(iz \cdot x)\, d\mu_k(z)$$

where the Q_k's are polynomials and the μ_k's are bounded measures supported by a certain finite family of algebraic subvarieties V_k of $V = \{z \in \mathbb{C}^n : P_1(z) = \dots = P_r(z) = 0\}$. The representation is a pointwise representation when the elements of X are actual functions (as it is the case with $X = \mathscr{O}$ or $X = \mathrm{Exp}$), while it has to be interpreted in the sense of generalized functions otherwise.

What is interesting from our point of view, however, is that the Fundamental Principle is an almost immediate consequence of what Ehrenpreis called the Structure Theorem, namely the existence of a topological isomorphism between the space $X^{\mathbf{P}}$ and the dual of a space of holomorphic functions satisfying suitable growth conditions on the union of the varieties V_k. The proof of the theorem is quite complicated, and even decades after Ehrenpreis' original announcement, it is still the subject of analysis and reinterpretations. The crux of the proof, however, is the understanding of the nature of the growth conditions that must be satisfied by holomorphic functions on the variety V for the duality to take place. Such growth conditions must be imposed not just on the functions, but also on suitable derivatives that somehow incorporate the multiplicities related to the polynomials P_j.

This said, in some cases this result can be remarkably simplified. For example, when $r = 1$, i.e. there is just a single differential equation, then one can consider the variety $V = \{z \in \mathbb{C}^n : P_1(z) = 0\}$, and if the variety does not have components with higher multiplicity than one, then the Q_k are constant polynomials, and the V_k are the algebraic varieties whose union is all of V. In this case, the definition of the space of holomorphic functions satisfying growth conditions on V turns out to be the obvious one, namely these are functions which are defined on V, and that satisfy (no request here on the derivatives) the expected growth.

This is clearly the case when we are considering, as a single differential operator, the complex Laplacian. Then the variety that Ehrenpreis' Fundamental Principle requires us to consider is exactly the complex light cone Γ, and since $\mathscr{O}$ and Exp are dual to each other, one can apply the Fundamental Principle (or, more properly, the Structure Theorem) to obtain directly both (1) and (2).

3 Complex Laplacian and Bicomplex Numbers

The study of the complex Laplacian is strictly intertwined with the theory of bicomplex and multicomplex numbers. Without giving many details (for which we refer the reader to the fairly comprehensive recent references [3, 7, 11, 12]), we recall that the set $\mathbb{BC}$ of bicomplex numbers is defined as the set of numbers of the form $Z = z_1 + jz_2$, where z_1 and z_2 are complex numbers, and j is an imaginary unit that commutes with the usual imaginary unit i. There is by now a very well developed (and quite interesting) theory for bicomplex valued functions defined on open sets in $\mathbb{BC}$, which can be represented as power series in Z. As it turns out, such functions are *bicomplex holomorphic*, in the sense that they admit *bicomplex derivative*, and satisfy a suitable system of first order differential equations, akin to the classical Cauchy-Riemann system (see [1] for the origins and first steps in this theory). In the theory of such functions, the complex light cone in two dimensions, i.e. $\Gamma = \{(z_1, z_2) : z_1^2 + z_2^2 = 0\}$, plays a very important role as it coincides with the set of zero-divisors in $\mathbb{BC}$. Similarly, the complex Laplacian plays also a prominent role because it can be factored as the product of two linear operators, one of which is the bicomplex differentiation.

As we mentioned before, bicomplex holomorphic functions are naturally connected to some sort of Cauchy-Riemann system. To be more precise, a function

$$f : U \subset \mathbb{BC} \to \mathbb{BC}$$

can obviously be written as $f(z_1, z_2) = u(z_1, z_2) + jv(z_1, z_2)$, and the condition of bicomplex holomorphicity on f is equivalent to the request that u and v are holomorphic functions of two variables, satisfying the additional Cauchy-Riemann like system

$$\frac{\partial u}{\partial z_1} = \frac{\partial v}{\partial z_2} \quad \text{and} \quad \frac{\partial v}{\partial z_1} = -\frac{\partial u}{\partial z_2}. \tag{3}$$

Thus, Ehrenpreis' Fundamental Principle immediately applies to bicomplex holomorphic functions, as they are solutions, in a LAU-space, of a system of linear constant coefficients partial differential equations. As such, one can give a suitable integral representation for bicomplex holomorphic functions, or regard them as suitable functionals on a space of functions of exponential type on the variety associated to the system (3), which not surprisingly turns out to be the complex light cone. As solutions of systems of linear constant coefficients differential equations, the space of bicomplex holomorphic function is, by itself, an LAU-space, and it would be interesting to develop the analog of the Fundamental Principle for bicomplex holomorphic solutions of differential equations. Indeed, one should even explore the possibility (just like in the case of holomorphic functions) of studying infinite order differential operators on the space of bicomplex holomorphic functions. It is well known that in the usual holomorphic case, such operators are nothing but analytic functionals carried by the origin, and as such they can be thought of as hyperfunctions with support in the origin, in view of the Martineau-Harvey theorem. Since a theory of bicomplex (and multicomplex) hyperfunctions has been studied as well,

[4, 15], it is natural to explore how this would play out in this context (again, we refer the reader to [13] where these ideas are developed further). One can imagine that the variety which would act as support for the integral representation for solutions of such differential operators would be the intersection of the complex light cone and the variety associated to the infinite order differential operator (such a variety would not be algebraic, so any such result would need to include a version of the Fundamental Principle in line with the ideas developed in [10]). I would like to conclude this section with a brief mention of what we could envision if we were to now consider functions defined on $\mathbb{BC}^2$, a space that is also referred to as the multicomplex space $\mathbb{BC}_3$. The elements of this space are pairs of bicomplex numbers $\mathscr{Z} = (Z_1, Z_2)$, which can also be thought of as multicomplex numbers $\mathscr{Z} = Z_1 + i_3 Z_2$, where i_1 and i_2 correspond to the two imaginary units i, j in $\mathbb{BC}$, and i_3 is a third imaginary unit that also commutes with the other two. Then one could consider the *bicomplex Laplacian* defined by

$$B\Delta = \frac{\partial}{\partial Z_1^2} + \frac{\partial}{\partial Z_2^2},$$

whose symbol is now $Z_1^2 + Z_2^2$. If we denote by Γ once again the bicomplex light cone obtained as the set $\Gamma = \{(Z_1, Z_2) \in \mathbb{BC}_3 : Z_1^2 + Z_2^2 = 0\}$, then one should be able to use the same ideas highlighted in the previous section to identify the strong dual of the set of multicomplex holomorphic functions on Γ with a suitable space of multicomplex holomorphic functions of exponential growth, satisfying the equation $B\Delta(F) = 0$.

References

1. Baley Price, G.: An Introduction to Multicomplex Spaces and Functions. Dekker, New York (1991)
2. Berenstein, C.A., Struppa, D.C.: Complex analysis and convolution equations. In: Several Complex Variables V. Encyclopaedia of Mathematical Sciences, vol. 54, pp. 1–109. Springer, Berlin (1993)
3. Colombo, F., Sabadini, I., Struppa, D.C., Vajiac, A., Vajiac, M.: Singularities of functions of one and several bicomplex functions. Ark. Mat. **49**, 277–294 (2011)
4. Colombo, F., Sabadini, I., Struppa, D.C., Vajiac, A., Vajiac, M.: Bicomplex hyperfunction theory. Ann. Mat. Pura Appl. **190**, 247–261 (2011)
5. Ehrenpreis, L.: The fundamental principle for linear constant coefficients partial differential equations. In: Proc. Int. Symp. Linear Spaces, Jerusalem, pp. 161–174 (1960)
6. Ehrenpreis, L.: Fourier Analysis in Several Complex Variables. Wiley-Interscience, New York (1970)
7. Luna-Elizarraras, M.E., Shapiro, M., Struppa, D.C., Vajiac, A.: Bicomplex numbers and their elementary functions. CUBO **14**, 61–80 (2012)
8. Morimoto, M., Fujita, K.: Analytic functionals and entire functionals on the complex light cone. Hiroshima Math. J. **25**, 493–512 (1995)
9. Palamodov, V.P.: Linear Differential Operators with Constant Coefficients. Springer, Berlin (1970)
10. Struppa, D.C.: The fundamental principle for systems of convolution equations. Mem. Am. Math. Soc. **273**, 1–167 (1983)

11. Struppa, D.C., Vajiac, A., Vajiac, M.: Remarks on holomorphicity in three settings: complex, quaternionic, and bicomplex. In: Hypercomplex Analysis and Applications. Trends in Mathematics, pp. 261–274. Birkhäuser, Basel (2010)
12. Struppa, D.C., Vajiac, A., Vajiac, M.: Holomorphy in multicomplex spaces. In: IWOTA 2010 Proceedings. Operator Theory: Advances and Applications, vol. 221, pp. 617–634. Birkhäuser, Basel (2012)
13. Struppa, D.C., Vajiac, A., Vajiac, M.: Ehrenpreis' fundamental principle and analytic functionals on the complex light cone (in preparation)
14. Treves, F.: Topological Vector Spaces, Distributions and Kernels. Academic Press, New York (1975)
15. Vajiac, A., Vajiac, M.: Multicomplex hyperfunctions. Complex Var. Elliptic Equ. **57**, 751–762 (2012)

The S-spectrum for Some Classes of Matrices

Mihaela B. Vajiac

Abstract In this paper we find the S-spectrum of operators given by particular classes of matrices, following the work of Colombo, Sabadini, Struppa and others. We treat the cases of general triangular matrices, some symmetric matrices and the special cases of Jacobi and resolvent matrices acting on respective Clifford algebras.

1 Introduction

The goal of this paper is to determine the S-spectrum of operators defined by certain real matrices $A_1, \dots, A_r$, with $r \geq 2$, and of some quaternionic matrices. In [17] and the many papers [3–6] preceding it, the authors set the basis for the S-functional calculus [8, 11, 12] and stressed the importance of computing values $f(A_1, \dots, A_r)$ for f a slice monogenic functions in a neighborhood of the S-spectrum of $A_1, \dots, A_r$. This paper aims to provide examples of calculus of the S-spectrum for some classes of matrices.

The S-functional calculus retains most of the properties of the Riesz-Dunford functional calculus for a single operator. We mention here that there exists a quaternionic version of the S-functional calculus, also called quaternionic functional calculus and, in its more general setting, can be found in [3, 4].

In the literature there exist other notions of spectra for r-tuples of operators, for example the joint spectrum see [20] or the monogenic spectrum, see [18, 19]. However, these notions of spectra differ from the S-spectrum since they are related to different classes of functions. The advantage of the use of the class of slice monogenic functions is that they allow a functional calculus for r-tuples of operators, commuting or non-commuting, bounded or unbounded, hence it can be viewed for the most general setting.

The theory of slice monogenic functions, mainly developed in the papers [2, 5, 10, 14–16] turned out to be very important because of its applications to these most general type of operators. The paper [9] introduced the notion of S-spectrum

M.B. Vajiac (✉)
Schmid College of Science and Technology, Chapman University, One University Drive, Orange, CA 92866, USA
e-mail: mbvajiac@chapman.edu

G. Gentili et al. (eds.), *Advances in Hypercomplex Analysis*, Springer INdAM Series 1, DOI 10.1007/978-88-470-2445-8_8, © Springer-Verlag Italia 2013

of r-tuples of not necessarily commuting operators, followed by a flurry of papers in this direction [6, 8, 11, 12]. Their work is related to the work of Jefferies and coauthors, see [18, 19] since it makes use of the Clifford algebra setting. This is a non-commutative setting, thus the computations are more complicated. The advantage of this type of functional calculus is that it allows to have a closed form for the resolvent operator.

The novelty of the present paper consists of the systematic treatment of the computation of S-spectrum for certain types of matrices. In particular we treat the cases of general triangular matrices, special symmetric matrices and the special cases of Jacobi, resolvent, and Hankel matrices acting on respective Clifford algebras.

We also point out that, in the case of commuting r-tuples of operators the computations of the S-spectrum can be usually simplified, by computing the F-spectrum instead (see [6, 7, 13]).

2 Notions of Functional Calculus

For consistency of the presentation, we give a brief introduction of functional calculus following the notations and definitions introduced in [17], as well as the main theorems of Functional Calculus. We denote by V a Banach space over $\mathbb{R}$ with norm $\|\cdot\|_V$ and we set $V_n := V \otimes \mathbb{R}_n$, where $\mathbb{R}_n$ is the real Clifford algebra. For an introduction to Clifford Analysis we refer the reader to [1]. We recall that V_n is a two-sided Banach module over $\mathbb{R}_n$ and its elements are of the type $\sum_A v_A \otimes e_A$, where $A = i_1 \dots i_r$, with $i_1 < \cdots < i_r$ is a multi-index, and $i_\ell \in \{1, 2, \dots, n\}$. The right and left multiplications of an element $v \in V_n$ with a scalar $a \in \mathbb{R}_n$ are defined as:

$$va = \sum_A v_A \otimes (e_A a)$$

and

$$av = \sum_A v_A \otimes (a e_A).$$

For simplicity of the presentation we will use the shortcut notation $\sum_A v_A e_A$ instead of $\sum_A v_A \otimes e_A$. We define the norm of elements in V_n by:

$$\|v\|^2_{V_n} := \sum_A \|v_A\|^2_V.$$

Let $\mathcal{B}(V)$ be the space of bounded $\mathbb{R}$-homomorphisms of the Banach space V into itself endowed with the natural norm denoted by $\|\cdot\|_{\mathcal{B}(V)}$. If $T_A \in \mathcal{B}(V)$, we can define the operator $T = \sum_A e_A T_A$ and its action on $v = \sum_B v_B e_B$ by:

$$T(v) = \sum_{A,B} T_A(v_B) e_A e_B.$$

The set of all such bounded operators is denoted by $\mathcal{B}_n(V_n)$ and the norm is defined by

$$\|T\|_{\mathcal{B}_n(V_n)} = \sum_A \|T_A\|_{\mathcal{B}(V)}.$$

We will omit the subscript $\mathcal{B}_n(V_n)$ in the norm of an operator when it is clear in context. Note also that $\|TS\| \leq \|T\|\|S\|$. A bounded operator $T = T_0 + \sum_{j=1}^n e_j T_j$, where $T_\mu \in \mathcal{B}(V)$ for $\mu = 0, 1, \dots, n$, will be called (abusing the terminology) an operator in paravector form. The set of such operators will be denoted by $\mathcal{B}_n^{0,1}(V_n)$. The set of bounded operators of the type $T = \sum_{j=1}^n e_j T_j$, where $T_\mu \in \mathcal{B}(V)$ for $\mu = 1, \dots, n$, will be denoted by $\mathcal{B}_n^1(V_n)$ and we will call T an operator in vector form.

There is a natural embedding of $\mathcal{B}_n^1(V_n)$ in $\mathcal{B}_n^{0,1}(V_n)$ and we will deal mostly with operators in vector form. The subset of the operators in $\mathcal{B}_n^1(V_n)$ whose components commute among themselves will be denoted by $\mathcal{BC}_n^1(V_n)$ and, similarly, we denote by $\mathcal{BC}_n^{0,1}(V_n)$ the set of paravector operators with commuting components.

We now recall some results introduced in [17]. Since we now want to construct a functional calculus for noncommuting operators we consider the noncommutative Cauchy kernel series in which we formally replace the paravector $\mathbf{x}$ by the paravector operator T, whose components do not necessarily commute. So we define the noncommutative Cauchy kernel operator series as follows.

Definition 1 Let $T \in \mathcal{B}_n^{0,1}(V_n)$ and $\mathbf{s} \in \mathbb{R}^{n+1}$. The $\mathcal{S}$-resolvent operator series is:

$$S^{-1}(\mathbf{s}, T) := \sum_{n\geq 0} T^n \mathbf{s}^{-1-n} \tag{1}$$

for $\|T\| < |\mathbf{s}|$, where $|\mathbf{s}|$ is the Euclidean norm of the paravector $\mathbf{s}$.

The most surprising fact is the following theorem that opens the way to the functional calculus for noncommuting operators.

Theorem 1 *Let $T \in \mathcal{B}_n^{0,1}(V_n)$ and $\mathbf{s} \in \mathbb{R}^{n+1}$. Then*

$$\sum_{n\geq 0} T^n \mathbf{s}^{-1-n} = -\big(T^2 - 2T\,\mathrm{Re}[\mathbf{s}] + |\mathbf{s}|^2\mathcal{I}\big)^{-1}(T - \bar{\mathbf{s}}\mathcal{I}), \tag{2}$$

for $\|T\| < |\mathbf{s}|$, where $\bar{\mathbf{s}}$ is the conjugate of the paravector $\mathbf{s}$.

The theorem above shows that even when the components of the operator $T = T_0 + e_1 T_1 + \cdots + e_n T_n$ do not necessarily commute, the sum of the series remains the same as in the case of a paravector $\mathbf{x}$. We can then define in a consistent way the $\mathcal{S}$-spectrum set as follows:

Definition 2 Let $T \in \mathcal{B}_n^{0,1}(V_n)$ and $\mathbf{s} \in \mathbb{R}^{n+1}$. We define the $\mathcal{S}$-*spectrum* of T by:

$$\sigma_S(T) := \big\{\mathbf{s} \in \mathbb{R}^{n+1} \mid T^2 - 2\,\mathrm{Re}[\mathbf{s}]T + |\mathbf{s}|^2\mathcal{I} \text{ is not invertible}\big\}.$$

The $\mathcal{S}$-*resolvent set* of T is defined by:

$$\rho_S(T) := \mathbb{R}^{n+1} \setminus \sigma_S(T).$$

Definition 3 Let $T \in \mathcal{B}_n^{0,1}(V_n)$ and $\mathbf{s} \in \rho_S(T)$. We define the $\mathcal{S}$-*resolvent operator* as:

$$S^{-1}(\mathbf{s}, T) := -\big(T^2 - 2\,\mathrm{Re}[\mathbf{s}]T + |\mathbf{s}|^2\mathcal{I}\big)^{-1}(T - \bar{\mathbf{s}}\mathcal{I}).$$

The $\mathcal{S}$-resolvent operator satisfies the following resolvent equation:

Theorem 2 *Let $T \in \mathcal{B}_n^{0,1}(V_n)$ and $\mathbf{s} \in \rho_S(T)$. Let $S^{-1}(\mathbf{s}, T)$ be the $\mathcal{S}$-resolvent operator. Then $S^{-1}(\mathbf{s}, T)$ satisfies the $\mathcal{S}$-resolvent equation*

$$S^{-1}(\mathbf{s}, T)\mathbf{s} - T S^{-1}(\mathbf{s}, T) = \mathcal{I}.$$

The following theorem gives the structure of the $\mathcal{S}$-spectrum.

Theorem 3 *Let $T \in \mathcal{B}_n^{0,1}(V_n)$ and suppose that $\mathbf{p} = p_0 + \underline{\mathbf{p}}$ belongs to $\sigma_S(T)$ with $\underline{\mathbf{p}} \neq 0$. Then all the elements of the $(n-1)$-sphere $[\mathbf{p}]$ belong to $\sigma_S(T)$.*

This result implies that if $\mathbf{p} \in \sigma_S(T)$ then either $\mathbf{p}$ is a real point or the whole $(n-1)$-sphere $[\mathbf{p}]$ belongs to $\sigma_S(T)$. For bounded paravector operators the $\mathcal{S}$-spectrum shares the same properties with the usual spectrum of a single operator, namely the spectrum is a compact and nonempty set:

Theorem 4 *Let $T \in \mathcal{B}_n^{0,1}(V_n)$. Then the $\mathcal{S}$-spectrum $\sigma_S(T)$ is a compact nonempty set. Moreover, $\sigma_S(T)$ is contained in $\{s \in \mathbb{R}^{n+1} \mid |\mathbf{s}| \leq \|T\|\}$.*

Without going into details, we recall the notion of $\mathcal{F}$-spectrum which was introduced for the first time in the paper [13], where the authors defined the $\mathcal{F}$-functional calculus. This is based on the integral version of the Fueter-Sce mapping theorem, and it associates to every slice monogenic function a function of operators $\breve{f}(T)$, where $\breve{f}$ is a subclass of the standard monogenic functions in the sense of Dirac. In this case the $\mathcal{F}$-spectrum is defined as follows:

Definition 4 Let $T \in \mathcal{B}_n^{0,1}(V_n)$, the $\mathcal{F}$-*spectrum* of T is defined as:

$$\sigma_F(T) = \big\{\mathbf{s} \in \mathbb{R}^{n+1} \,\big|\, \mathbf{s}^2 - \mathbf{s}(T + \bar{T}) + T\bar{T} \text{ is not invertible}\big\}.$$

For example in dimension 2, if $T = T_0 + T_1 e_1 + T_2 e_2$, then $\bar{T} = T_0 - T_1 e_1 - T_2 e_2$, $T + \bar{T} = 2T_0$, and $T\bar{T} = T_0^2 + T_1^2 + T_2^2$. In general, if T has commuting components, it is important to note that $\sigma_S(T) = \sigma_F(T)$, since the latter is easier to calculate.

More details of the $\mathcal{F}$-spectrum can be found in [7, 13].

The remainder of the paper is structured in several sections where we compute the $\mathcal{S}$-spectrum for several types of operators given by certain types of matrices. We begin with the operator $T = \sum_{i=1}^n A_i e_i$ given by $k \times k$ real triangular matrices $A_1, \ldots, A_n$. This operator is acting on $(\mathbb{R}_n)^k$ in the usual way.

3 Triangular Matrices as Operators on $\mathbb{R}_n$

3.1 Two 2 × 2 Triangular Matrices

As a warm-up exercise we start with 2×2 triangular matrices on $\mathbb{R}_2$, which yields insights on resolving the general case. However, this case is simplified by the fact that elements in $\mathbb{R}_2$ are invertible. Consider the operator $T = A_1e_1 + A_2e_2$, with $A_i = \left[\begin{smallmatrix} a_i & b_i \\ 0 & c_i \end{smallmatrix}\right]$, where all $a_i, b_i, c_i \in \mathbb{R}$. Then

$$T^2 - 2\,\mathrm{Re}[\mathbf{s}]T + |\mathbf{s}|^2 I = \begin{bmatrix} \alpha & \beta \\ 0 & \delta \end{bmatrix},$$

which is not invertible if and only if there exist a non-zero element $v = \left[\begin{smallmatrix} v_1 \\ v_2 \end{smallmatrix}\right]$, where $v_i \in \mathbb{R}_2$, such that

$$\begin{bmatrix} \alpha & \beta \\ 0 & \delta \end{bmatrix} \begin{bmatrix} v_1 \\ v_2 \end{bmatrix} = \begin{bmatrix} 0 \\ 0 \end{bmatrix}.$$

This is equivalent to $\alpha v_1 + \beta v_2 = \delta v_2 = 0$. Separating v_i and α, β, δ in components, we obtain:

$$\begin{aligned} v_i &= v_i^0 + v_i^1 e_1 + v_i^2 e_2 + v_i^3 e_1 e_2, \\ \alpha &= -a_1^2 - a_2^2 - 2s_0(a_1e_1 + a_2e_2) + |s|^2, \\ \beta &= -b_1(a_1 + c_1) - b_2(a_2 + c_2) + \big(b_2(a_1 - c_1) + b_1(c_2 - a_2)\big)e_1e_2 \\ &\quad - 2s_0(b_1e_1 + b_2e_2), \\ \delta &= -c_1^2 - c_2^2 - 2s_0(c_1e_1 + c_2e_2) + |s|^2. \end{aligned}$$

From the equation $\delta v_2 = 0$ we have:

$$\big(-c_1^2 - c_2^2 - 2s_0(c_1e_1 + c_2e_2) + |\mathbf{s}|^2\big)v_2 = 0,$$

yielding two cases: $v_2 = 0$, or $-c_1^2 - c_2^2 - 2s_0(c_1e_1 + c_2e_2) + |\mathbf{s}|^2 = 0$.

In the second case, we have $s_0 = 0$ and $s_1^2 + s_2^2 = c_1^2 + c_2^2$ or the special case $s_0 \neq 0$ and $c_1 = c_2 = 0$.

In the first case, if $v_2 = 0$ then $Av_1 = 0$ and $v_1 \neq 0$, therefore

$$-a_1^2 - a_2^2 - 2s_0(a_1e_1 + a_2e_2) + |\mathbf{s}|^2 = 0.$$

Thus $s_0 = 0$ and $s_1^2 + s_2^2 = a_1^2 + a_2^2$ or the special case $s_0 \neq 0$ and $a_1 = a_2 = 0$.

Summarizing, we obtain the following result:

Theorem 5 *The operator $T = A_1e_1 + A_2e_2$, has the following spherical spectrum:*

$$\sigma_S(T) = \mathcal{S} = \big\{\mathbf{s} \in \mathbb{R}^3 \mid s_0 = 0,\ |\mathbf{s}|^2 = c_1^2 + c_2^2 \text{ or } |\mathbf{s}|^2 = a_1^2 + a_2^2\big\}.$$

3.2 Three 2 × 2 Triangular Matrices

For three 2×2 triangular matrices we build the operator $T = A_1e_1 + A_2e_2 + A_3e_3$, acting on vectors with elements in $\mathbb{R}_3$. As before, if $A_i = \left[\begin{smallmatrix} a_i & b_i \\ 0 & c_i \end{smallmatrix}\right]$, where a_i, b_i, c_i, are real numbers we have:

$$T^2 - 2\,\mathrm{Re}[\mathbf{s}]T + |\mathbf{s}|^2 I = \begin{bmatrix} \alpha & \beta \\ 0 & \delta \end{bmatrix},$$

where:

$$\alpha = -a_1^2 - a_2^2 - a_3^2 - 2s_0(a_1e_1 + a_2e_2) + |s|^2,$$

$$\beta = \sum_{i<j}(a_ib_j - a_jb_i + b_ic_j - b_jc_i)e_ie_j - \sum_{i=1}^{3}(a_ib_i + b_ic_i) - 2s_0\sum_{i=1}^{3} b_ie_i,$$

$$\delta = -c_1^2 - c_2^2 - c_3^2 - 2s_0(c_1e_1 + c_2e_2 + c_3e_3) + |\mathbf{s}|^2.$$

Then $T^2 - 2\,\mathrm{Re}[\mathbf{s}]T + |\mathbf{s}|^2 I$ is not invertible if and only if there exist a non-zero element $v = \left[\begin{smallmatrix} v_1 \\ v_2 \end{smallmatrix}\right]$, where $v_i \in \mathbb{R}_3$, such that

$$\begin{bmatrix} \alpha & \beta \\ 0 & \delta \end{bmatrix}\begin{bmatrix} v_1 \\ v_2 \end{bmatrix} = \begin{bmatrix} 0 \\ 0 \end{bmatrix},$$

equivalently $\alpha v_1 + \beta v_2 = \delta v_2 = 0$, where $v_i \in \mathbb{R}_3$, hence not invertible anymore.

However, in the last equation $\delta v_2 = 0$, δ is a paravector, therefore invertible, so this equation will yield either $\delta = 0$ or $v_2 = 0$. If $v_2 = 0$ then the first equation becomes $\alpha v_1 = 0$, and we obtain a similar characterization of the $\mathcal{S}$-spectrum as follows:

Theorem 6 *Let $T = A_1e_1 + A_2e_2 + A_3e_3$, where A_i are triangular matrices as above, then the spectrum is given by:*

$$\begin{aligned}\sigma_S(T) &= \{\mathbf{s} \in \mathbb{R}^3 \mid \delta = 0 \text{ or } \alpha = 0\} \\ &= \{\mathbf{s} \in \mathbb{R}^3 \mid s_0 = 0,\ |\mathbf{s}|^2 = c_1^2 + c_2^2 + c_3^2 \text{ or } |\mathbf{s}|^2 = a_1^2 + a_2^2 + a_3^2\}.\end{aligned}$$

3.3 k × k Triangular Matrices

For the general case of $k \times k$ real triangular matrices which build an operator $T = \sum_{i=1}^{n} A_ie_i$ on $\mathbb{R}_n$, we can use the methods of the previous example. The arguments are similar, as the components of the resulting matrix $T^2 - 2\,\mathrm{Re}[\mathbf{s}]T + |\mathbf{s}|^2 I$ will be paravectors, even if the matrix acts on $\mathbb{R}_n$. Consider $T = \sum_{i-1}^{n} A_ie_i$ where

$$A_i = \begin{pmatrix} a_1^i & a_{12}^i & a_{13}^i & \cdots & a_{1k}^i \\ 0 & a_2^i & a_{23}^i & \cdots & a_{2k}^i \\ \vdots & \vdots & \vdots & \vdots & \vdots \\ 0 & 0 & 0 & \cdots & a_k^i \end{pmatrix}.$$

Then the resolvent $T^2 - 2\operatorname{Re}[\mathbf{s}]T + |\mathbf{s}|^2 I$ is given by the following matrix:

$$\begin{pmatrix} \Omega_1 & \Gamma_{12} & \Gamma_{13} & \cdots & \Gamma_{1k-1} \\ 0 & \Omega_2 & \Gamma_{23} & \cdots & \Gamma_{2k-1} \\ \vdots & \vdots & \vdots & \vdots & \vdots \\ 0 & 0 & 0 & \cdots & \Omega_k \end{pmatrix},$$

where

$$\Omega_j := -\sum_{i=1}^{n} (a_j^i)^2 - 2s_0 \sum_{i=1}^{n} a_i e_i + |\mathbf{s}|^2,$$

for all $1 \le j \le k$ are paravectors, as well as all other Γ_{ij}.

In the same fashion as the 2×2 case, we analyze the system given by $(T^2 - 2\operatorname{Re}[\mathbf{s}]T + |\mathbf{s}|^2 I)V = 0$, where V is a column of elements $v_i \in \mathbb{R}_n$. The last equation is

$$\Omega_k v_k = 0,$$

and, since

$$\Omega_k = -\sum_{i=1}^{n} (a_k^i)^2 - 2s_0 \sum_{i=1}^{n} a_i e_i + |\mathbf{s}|^2$$

is a paravector, it yields either

$$-\sum_{i=1}^{n} (a_k^i)^2 - 2s_0 \sum a_i e_i + |\mathbf{s}|^2 = 0,$$

or $v_k = 0$. The discussion continues in the same fashion, and we obtain the following general characterization of the spectrum of this type of operator:

Theorem 7 *Consider $A_1, \dots, A_n$ $k \times k$ triangular matrices and the operator $T = \sum_{i=1}^{n} A_i e_i$ acting on columns of elements in $\mathbb{R}_n$. Then the spectrum of the operator T is:*

$$\sigma_S(T) = \left\{ \mathbf{s} \in \mathbb{R}^{n+1} \mid s_0 = 0,\ |\mathbf{s}|^2 = \sum_{i=1}^{n} (a_1^i)^2,\ or \dots,\ or\ |\mathbf{s}|^2 = \sum_{i=1}^{n} (a_k^i)^2 \right\}.$$

4 Symmetric Matrices

4.1 Special Symmetric Matrices

We will treat here the case of operators obtained from special symmetric matrices $A_i = \left[\begin{smallmatrix} a_i & b_i \\ b_i & a_i \end{smallmatrix}\right]$. For $T = \sum_{i=1}^{3} A_i e_i$, the resolvent operator becomes:

$$T^2 - 2\operatorname{Re}[\mathbf{s}]T + |\mathbf{s}|^2 I = -\sum_{i=1}^{3} A_i^2 - 2s_0 \sum_{i=1}^{3} A_i e_i + |\mathbf{s}|^2 I,$$

and we analyze the system $(T^2 - 2\,\mathrm{Re}[\mathbf{s}]T + |\mathbf{s}|^2 I)V = 0$, where the components of V are $v_i = q_{i1} + e_3 q_{i2}$, elements of $\mathbb{R}_3$. The system is equivalent to:

$$Cv_1 + Dv_2 = Dv_1 + Cv_2 = 0,$$

where:

$$C = -\sum_{i=1}^{3}\big((a_i)^2 + (b_i)^2\big) + |\mathbf{s}|^2 - 2s_0 \sum_{i=1}^{3} a_i e_i$$

and

$$D = -2\sum_{i=1}^{3} a_i b_i - 2s_0 \sum_{i=1}^{3} b_i e_i$$

are paravectors, therefore invertible if non-zero.

The spectrum will then contain the set: $\{s \in \mathbb{R}^4 \mid s_0 = 0,\ |\mathbf{s}|^2 = \sum_{i=1}^{3}(a_i)^2 + (b_i)^2\}$, predicting a more complicated spectrum for general symmetric matrices, which are treated below.

4.2 General Symmetric Matrices

The case of operators obtained from general symmetric matrices becomes difficult even for 2×2 matrices. If we let $T = A_1 e_1 + A_2 e_2 + A_3 e_3$, with $A_i = \left[\begin{smallmatrix} a_i & b_i \\ b_i & c_i \end{smallmatrix}\right]$ symmetric matrices of real numbers, we have:

$$T^2 - 2\,\mathrm{Re}[\mathbf{s}]T + |\mathbf{s}|^2 I = \begin{bmatrix} \alpha & \beta \\ \gamma & \delta \end{bmatrix},$$

where:

$$\alpha = -\sum_{i=1}^{3} a_i^2 - \sum_{i=1}^{3} b_i^2 - 2s_0 \sum_{i=1}^{3} a_i e_i + |\mathbf{s}|^2,$$

$$\beta = \sum_{i<j}(a_i b_j - a_j b_i + b_i c_j - b_j c_i)e_i e_j - \sum_{i=1}^{3}(a_i b_i + b_i c_i) - 2s_0 \sum_{i=1}^{3} b_i e_i,$$

$$\gamma = -\sum_{i<j}(a_i b_j - a_j b_i + b_i c_j - b_j c_i)e_i e_j - \sum_{i=1}^{3}(a_i b_i + b_i c_i) - 2s_0 \sum_{i=1}^{3} b_i e_i,$$

$$\delta = -\sum_{i=1}^{3} b_i^2 - \sum_{i=1}^{3} c_i^2 - 2s_0 \sum_{i=1}^{3} c_i e_i + |\mathbf{s}|^2.$$

Again, to find the S-spectrum of T we have to find conditions for which the system $(T^2 - 2\,\mathrm{Re}[\mathbf{s}]T + |\mathbf{s}|^2 I)V = 0$, has a non-degenerate solution, where the components $v_i = q_{i1} + e_3 q_{i2}$ of V are elements of $\mathbb{R}_3$. The system becomes $\alpha v_1 + \beta v_2 = \gamma v_1 + \delta v_2 = 0$.

Consider the case $\delta = 0$, then $s_0 = 0$ and $|\mathbf{s}|^2 = \sum_{i=1}^3 b_i^2 + \sum_{i=1}^3 c_i^2$, then

$$\alpha = \sum_{i=1}^{3} c_i^2 - \sum_{i=1}^{3} a_i^2,$$

$$\beta = \sum_{i<j} (a_i b_j - a_j b_i + b_i c_j - b_j c_i) e_i e_j - \sum_{i=1}^{3} (a_i b_i + b_i c_i),$$

$$\gamma = -\sum_{i<j} (a_i b_j - a_j b_i + b_i c_j - b_j c_i) e_i e_j - \sum_{i=1}^{3} (a_i b_i + b_i c_i),$$

$$\delta = 0,$$

and the last equation of the system $(T^2 - 2\,\mathrm{Re}[\mathbf{s}]T + |\mathbf{s}|^2 I)V = 0$, with the components $v_i = q_{i1} + e_3 q_{i2}$ of V in $\mathbb{R}_3$, becomes $\gamma v_1 = 0$.

Rewriting

$$\Lambda = \sum (a_i b_i + b_i c_i),$$

$$\Phi = (a_1 b_3 - a_3 b_1 + b_1 c_3 - b_3 c_1) e_1 + (a_2 b_3 - a_3 b_2 + b_2 c_3 - b_3 c_2) e_2,$$

$$\Psi = (a_1 b_2 - a_2 b_1 + b_1 c_2 - b_2 c_1) e_1 e_2,$$

and separating the equation $\gamma v_1 = 0$, in its components, we obtain:

$$-\Lambda q_{11} + \Phi q_{12} - \Psi q_{11} = 0,$$

$$-\Lambda q_{12} - \Phi q_{11} - \Psi q_{12} = 0.$$

If $\Phi = 0$ then both $q_{11} = q_{12} = 0$, otherwise, since Φ is a paravector, it will have an inverse $\Phi^{-1} = -\frac{1}{|\Phi|^2}\Phi$ and we can solve the system as follows: $q_{22} = \Phi^{-1}(\Lambda + \Psi) q_{21}$ and $(-\Phi - (\Lambda + \Psi)\Phi^{-1}(\Lambda + \Psi)) q_{21} = 0$. So, if q_{21} and q_{22} are non-zero then $-\Phi - (\Lambda + \Psi)\Phi^{-1}(\Lambda + \Psi) = 0$, or $\Phi|\Phi|^2 = (\Lambda + \Psi)\Phi(\Lambda + \Psi)$.

We obtain an equivalent case when $\alpha = 0$, and the spectrum will contain the sets: $\{s \in \mathbb{R}^4 \mid \alpha = 0\}$ and $\{s \in \mathbb{R}^4 \mid \delta = 0\}$, or, equivalently, $\{s \in \mathbb{R}^4 \mid s_0 = 0,\ |\mathbf{s}|^2 = \sum_{i=1}^3 (a_i)^2 + (b_i)^2\}$ and $\{s \in \mathbb{R}^4 \mid s_0 = 0,\ |\mathbf{s}|^2 = \sum_{i=1}^3 (b_i)^2 + (c_i)^2\}$.

We should also note that for an operator formed by two symmetric matrices $T = A_1 e_1 + A_2 e_2$ this becomes a special case of Hankel matrix treated in the next section.

5 Other Special Matrices

5.1 The Jordan Matrix

We consider the operator given by the Jordan matrix in the quaternionic ($\mathbb{H} = \mathbb{R}_2$) case and calculate its spectrum as follows.

Let $T = J_m(q)$ the Jordan matrix given by:

$$J_m(q) = \begin{pmatrix} q & 1 & 0 & \cdots & 0 \\ 0 & q & 1 & \cdots & 0 \\ \vdots & \vdots & \vdots & \vdots & \vdots \\ 0 & 0 & 0 & \cdots & q \end{pmatrix},$$

where $q \in \mathbb{H}$. To simplify this operator, we decompose it: $T = qI_n + A_n$, where the matrix A_n is:

$$A_n = \begin{pmatrix} 0 & 1 & 0 & \cdots & 0 \\ 0 & 0 & 1 & \cdots & 0 \\ \vdots & \vdots & \vdots & \vdots & \vdots \\ 0 & 0 & 0 & \cdots & 0 \end{pmatrix}.$$

Then, we have that $T^2 = q^2 I_n + 2qA + B_n$, where the matrix B_n is given by:

$$B_n = \begin{pmatrix} 0 & 0 & 1 & 0 \cdots & 0 \\ 0 & 0 & 0 & 1 \cdots & 0 \\ \vdots & \vdots & \vdots & \vdots & \vdots \\ 0 & 0 & 0 & 0 \cdots & 0 \end{pmatrix}.$$

Therefore we have:

$$T^2 - 2\,\mathrm{Re}[s]T + |s|^2 I = \left(q^2 - 2s_0 q + |s|^2\right) I_n + (2q - 2s_0 q)A_n + B_n.$$

Rewriting $\alpha = q^2 - 2s_0 q + |s|^2$ and $\beta = 2q - 2s_0 q$, then the system

$$\left(T^2 - 2\,\mathrm{Re}[s]T + |s|^2 I\right) Q = 0,$$

(where Q is a column of quaternions $q_i \in \mathbb{H}$) becomes:

$$\begin{aligned} \alpha q_1 + \beta q_2 + q_3 &= 0, \\ \alpha q_2 + \beta q_3 + q_4 &= 0, \\ &\vdots \\ \alpha q_{n-2} + \beta q_{n-1} + q_n &= 0, \\ \alpha q_{n-1} + \beta q_n &= 0, \\ \alpha q_n &= 0. \end{aligned}$$

Working backwards from the last equation, we have that $\alpha = 0$ or $q_n = 0$. If $q_n = 0$ then the previous equation becomes $\alpha q_{n-1} = 0$ and so forth, and we obtain the following representation of the S-spectrum of T:

Theorem 8 *The spectrum of the operator $T = J_m(q)$, where $J_m(q)$ is the Jordan matrix*

$$J_m(q) = \begin{pmatrix} q & 1 & 0 & \cdots & 0 \\ 0 & q & 1 & \cdots & 0 \\ \vdots & \vdots & \vdots & \vdots & \vdots \\ 0 & 0 & 0 & \cdots & q \end{pmatrix},$$

for any $q \in \mathbb{H}$, is given by:

$$\sigma_S(T) = \{s \in \mathbb{H} \mid \alpha = 0\} = \left\{s \in \mathbb{H} \mid q^2 - 2s_0 q + |s|^2 = 0\right\}.$$

This spectrum coincides with the sphere defined by q, denoted by $[q]$.

5.2 A Special Resolvent Matrix

We will remain for a while longer in the quaternionic case $\mathbb{H} = \mathbb{R}_2$, and we will consider the operator given by a special resolvent matrix:

$$R_n(q) = \begin{pmatrix} 1 & 0 & \cdots & 0 \\ q & 1 & \cdots & 0 \\ \vdots & \vdots & \vdots & \vdots \\ q^n & q^{n-1} & \cdots & 1 \end{pmatrix},$$

for $q \in \mathbb{H}$ a fixed quaternion. Just as in the previous case we decompose: $T = I_n + A_n$, where the matrix A_n is:

$$A_n = \begin{pmatrix} 0 & 0 & 0 & \cdots & 0 & 0 \\ q & 0 & 0 & \cdots & 0 & 0 \\ \vdots & \vdots & \vdots & \vdots & \vdots & \vdots \\ q^{n-1} & q^{n-2} & q^{n-3} & \cdots & q & 0 \end{pmatrix},$$

and we obtain $T^2 = I_n + 2A_n + A_n^2$. A quick computation yields the matrix:

$$A_n^2 = \begin{pmatrix} 0 & 0 & 0 & \cdots & 0 & 0 & 0 \\ 0 & 0 & 0 & \cdots & 0 & 0 & 0 \\ q^2 & 0 & 0 & \cdots & 0 & 0 & 0 \\ 2q^3 & q^2 & 0 & \cdots & 0 & 0 & 0 \\ \vdots & \vdots & \vdots & \vdots & \vdots & \vdots & \vdots \\ (n-2)q^{n-1} & (n-3)q^{n-2} & (n-4)q^{n-3} & \cdots & q^2 & 0 & 0 \end{pmatrix},$$

therefore:

$$T^2 - 2\,\mathrm{Re}[s]T + |s|^2 I = \left(1 - 2s_0 + |s|^2\right) I_n + (2 - 2s_0) A_n + A_n^2.$$

Rewriting $\alpha = 1 - 2s_0 + |s|^2$ and $\beta_k = (k + 1 - 2s_0)q^k$, $1 \le k \le n-1$ the system $(T^2 - 2\,\mathrm{Re}[s]T + |s|^2 I)Q = 0$ (where Q is a column of quaternions $q_i \in \mathbb{H}$) becomes:

$$
\begin{aligned}
\alpha q_1 &= 0,\\
\beta_1 q_1 + \alpha q_2 &= 0,\\
\beta_2 q_1 + \beta_1 q_2 + \alpha q_3 &= 0,\\
\beta_3 q_1 + \beta_2 q_2 + \beta_1 q_3 + \alpha q_4 &= 0,\\
&\vdots\\
\beta_{n-2} q_1 + \beta_{n-3} q_2 + \cdots + \beta_1 q_{n-2} + \alpha q_{n-1} &= 0,\\
\beta_{n-1} q_1 + \beta_{n-2} q_2 + \cdots + \beta_1 q_{n-1} + \alpha q_n &= 0.
\end{aligned}
$$

It is easy to see that $\alpha = 0$ or $q_1 = 0$. If $q_1 = 0$ then the next equation becomes $\alpha q_2 = 0$ and so forth, so we obtain the following representation of the S-spectrum of T:

Theorem 9 *The spectrum of the operator $T = R_n(q)$, where $R_n(q)$ is the resolvent matrix*

$$
R_n(q) = \begin{pmatrix} 1 & 0 & \cdots & 0 \\ q & 1 & \cdots & 0 \\ \vdots & \vdots & \vdots & \vdots \\ q^n & q^{n-1} & \cdots & 1 \end{pmatrix},
$$

for any $q \in \mathbb{H}$, is given by:

$$
\sigma_S(T) = \{s \in \mathbb{H} \mid \alpha = 0\} = \{s \in \mathbb{H} \mid 1 - 2s_0 + |s|^2 = 0\}.
$$

Reordering the terms above, this spectrum coincides with the real point 1.

5.3 The Hankel Matrix

We now turn to the operator given by the Hankel matrix:

$$
\Gamma_m = \begin{pmatrix} \mu_0 & \mu_1 & \cdots & \mu_{m-1} \\ \mu_1 & \mu_2 & \cdots & \mu_m \\ \vdots & \vdots & \vdots & \vdots \\ \mu_{m-1} & \mu_m & \cdots & \mu_{2m-2} \end{pmatrix},
$$

where $\mu_j \in \mathbb{H}$. This case has not been solved yet in full generality, but in the case $m = 2$ we have a representation of the spectrum. In this case the matrix is a 2×2 symmetric matrix of quaternions:

$$
\Gamma_2 = \begin{pmatrix} \mu_0 & \mu_1 \\ \mu_1 & \mu_2 \end{pmatrix},
$$

where $\mu_0, \mu_1, \mu_2 \in \mathbb{H}$. For $T = \Gamma_2$, we obtain the resolvent operator:

$$T^2 - 2\,\mathrm{Re}[\mathbf{s}]T + |\mathbf{s}|^2 I = \begin{pmatrix} \mu_0^2 + \mu_1^2 - 2s_0\mu_0 + |\mathbf{s}|^2 & \mu_0\mu_1 + \mu_1\mu_2 - 2s_0\mu_1 \\ \mu_1\mu_0 + \mu_2\mu_1 - 2s_0\mu_1 & \mu_2^2 + \mu_1^2 - 2s_0\mu_2 + |\mathbf{s}|^2 \end{pmatrix}.$$

Rewriting $T^2 - 2\,\mathrm{Re}[\mathbf{s}]T + |\mathbf{s}|^2 I = \left(\begin{smallmatrix} A & B \\ C & D \end{smallmatrix}\right)$, the system

$$\left(T^2 - 2\,\mathrm{Re}[\mathbf{s}]T + |\mathbf{s}|^2 I\right)Q = 0$$

becomes: $Aq_1 + Bq_2 = Cq_1 + Dq_2 = 0$.

If either $q_1 = 0$ or $q_2 = 0$ we obtain $A = C = 0$ or, equivalently $B = D = 0$ and the spectrum given by

$$\mu_0^2 + \mu_1^2 - 2s_0\mu_0 + |\mathbf{s}|^2 = \mu_1\mu_0 + \mu_2\mu_1 - 2s_0\mu_1 = 0,$$

or

$$\mu_0\mu_1 + \mu_1\mu_2 - 2s_0\mu_1 = \mu_2^2 + \mu_1^2 - 2s_0\mu_2 + |\mathbf{s}|^2 = 0.$$

If neither $q_1 = 0$ nor $q_2 = 0$, then the spectrum is given by either $AD^{-1}C + B = 0$ or $DB^{-1}A + C = 0$. The final form of the characterization of this spectrum is quite messy and we choose to leave it in this form. The general Hankel case should prove challenging and will become the object of further study.

As an example of this computation we offer the particular case of the Hankel matrix:

$$\Gamma_2 = \begin{pmatrix} e_1 & e_2 \\ e_2 & e_1e_2 \end{pmatrix},$$

the S-spectrum is given by: $\{s \in \mathbb{R}^3 \mid s_0 = 0,\ |\mathbf{s}|^2 = 2 \pm \sqrt{2}\}$, which we leave as an exercise to the reader.

5.4 Pauli Matrices

In the case of two 2×2 Pauli matrices the S-spectrum and the monogenic spectrum have been calculated in [17]. The case of three 4×4 Pauli matrices yielding the operator $T = \sigma_1 e_1 + \sigma_2 e_2 + \sigma_3 e_3$ on $\mathbb{R}_3$ proves more challenging, where

$$\sigma_1 = \begin{pmatrix} 0 & 0 & 1 & 0 \\ 0 & 0 & 0 & 1 \\ 1 & 0 & 0 & 0 \\ 0 & 1 & 0 & 0 \end{pmatrix}, \qquad \sigma_2 = \begin{pmatrix} 0 & 0 & 0 & -1 \\ 0 & 0 & 1 & 0 \\ 0 & 1 & 0 & 0 \\ -1 & 0 & 0 & 0 \end{pmatrix},$$

$$\sigma_3 = \begin{pmatrix} 1 & 0 & 0 & 0 \\ 0 & 1 & 0 & 0 \\ 0 & 0 & -1 & 0 \\ 0 & 0 & 0 & -1 \end{pmatrix},$$

are the three 4×4 Pauli matrices. A simple step yields:

$$T = \begin{pmatrix} e_3 & 0 & e_1 & -e_2 \\ 0 & e_3 & e_2 & e_1 \\ e_1 & e_2 & e_3 & 0 \\ -e_2 & e_1 & 0 & e_3 \end{pmatrix},$$

followed by:

$$T^2 = \begin{pmatrix} -3 & 2e_1e_2 & 2e_3e_1 & 2e_2e_3 \\ 2e_2e_1 & -3 & 2e_2e_3 & 2e_3e_1 \\ 2e_1e_3 & 2e_2e_3 & -3 & 2e_2e_1 \\ 2e_3e_2 & 2e_1e_3 & 2e_1e_2 & -3 \end{pmatrix},$$

equivalently:

$$T^2 = \begin{pmatrix} -3 & 2e_1e_2 & -2e_1e_3 & 2e_2e_3 \\ -2e_1e_2 & -3 & 2e_2e_3 & -2e_1e_3 \\ 2e_1e_3 & 2e_2e_3 & -3 & -2e_1e_2 \\ -2e_2e_3 & 2e_1e_3 & 2e_1e_2 & -3 \end{pmatrix}.$$

Then the matrix for the resolvent operator $T^2 - 2s_0T + |\mathbf{s}|^2 I$ becomes:

$$\begin{pmatrix} -3-2s_0e_3+|\mathbf{s}|^2 & 2e_1e_2 & 2e_3e_1-2s_0e_1 & 2e_2e_3+2s_0e_2 \\ 2e_2e_1 & -3-2s_0e_3+|\mathbf{s}|^2 & 2e_2e_3-2s_0e_2 & 2e_3e_1-2s_0e_1 \\ 2e_1e_3-2s_0e_1 & 2e_2e_3-2s_0e_2 & -3+2s_0e_3+|\mathbf{s}|^2 & 2e_2e_1 \\ 2e_3e_2+2s_0e_2 & 2e_1e_3-2s_0e_1 & 2e_1e_2 & -3+2s_0e_3+|\mathbf{s}|^2 \end{pmatrix},$$

and we apply this to a column of elements in $\mathbb{R}_3$.

Using Computer Algebra Systems (e.g. Maple) we were able to derive the solution to the spectrum of this operator, albeit a long and intricate formula that, for the sake of the presentation, we will not write in this present paper. We will just mention that the spectrum of this operator includes the set: $\{\mathbf{s} \in \mathbb{R}^4 \mid s_0 = 0 \text{ and } |\mathbf{s}|^2 = 3\}$.

References

1. Brackx, F., Delanghe, R., Sommen, F.: Clifford Analysis. Pitman Research Notes in Mathematics, vol. 76 (1982)
2. Colombo, F., Sabadini, I.: A structure formula for slice monogenic functions and some of its consequences. In: Hypercomplex Analysis. Trends in Mathematics, pp. 69–99. Birkhäuser, Basel (2009)
3. Colombo, F., Sabadini, I.: On some properties of the quaternionic functional calculus. J. Geom. Anal. **19**, 601–627 (2009)
4. Colombo, F., Sabadini, I.: On the formulations of the quaternionic functional calculus. J. Geom. Phys. **60**, 1490–1508 (2010)
5. Colombo, F., Sabadini, I.: The Cauchy formula with s-monogenic kernel and a functional calculus for noncommuting operators. J. Math. Anal. Appl. **373**, 655–679 (2011)
6. Colombo, F., Sabadini, I.: Some remarks on the S-spectrum. Complex Var. Elliptic Equ., 1–6 (2011, iFirst)

7. Colombo, F., Sabadini, I.: The $\mathcal{F}$-spectrum and the $\mathcal{SC}$-functional calculus. Proc. R. Soc. Edinb., Sect. A, Math. **142**(3), 479–500 (2012)
8. Colombo, F., Gentili, G., Sabadini, I., Struppa, D.: A functional calculus in an non-commutative setting. Electron. Res. Announc. Math. Sci. **14**, 60–68 (2007)
9. Colombo, F., Sabadini, I., Struppa, D.C.: A new functional calculus for noncommuting operators. J. Funct. Anal. **254**, 2255–2274 (2008)
10. Colombo, F., Sabadini, I., Struppa, D.C.: Slice monogenic functions. Isr. J. Math. **171**, 385–403 (2009)
11. Colombo, F., Gentili, G., Sabadini, I., Struppa, D.: Non-commutative functional calculus, bounded operators. Complex Anal. Oper. Theory **4**(4), 841–843 (2010)
12. Colombo, F., Gentili, G., Sabadini, I., Struppa, D.: Non-commutative functional calculus, unbounded operators. J. Geom. Phys. **60**, 251–259 (2010)
13. Colombo, F., Sabadini, I., Sommen, F.: The Fueter mapping theorem in integral form and the $\mathcal{F}$-functional calculus. Math. Methods Appl. Sci. **33**, 2050–2066 (2010)
14. Colombo, F., Sabadini, I., Struppa, D.C.: Extension properties for slice monogenic functions. Isr. J. Math. **177**, 369–389 (2010)
15. Colombo, F., Sabadini, I., Struppa, D.C.: Duality theorems for slice hyperholomorphic functions. J. Reine Angew. Math. **645**, 85–104 (2010)
16. Colombo, F., Sabadini, I., Struppa, D.C.: The Runge theorem for slice hyperholomorphic functions. Proc. Am. Math. Soc. **139**, 1787–1803 (2011)
17. Colombo, F., Sabadini, I., Struppa, D.C.: Noncommutative Functional Calculus. Theory and Applications of Slice Hyperholomorphic Functions. Progress in Mathematics, vol. 289. Birkhäuser, Basel (2011)
18. Jefferies, B.: Spectral Properties of Noncommuting Operators. Lecture Notes in Mathematics, vol. 1843. Springer, Berlin (2004)
19. Jefferies, B., McIntosh, A., Picton-Warlow, J.: The monogenic functional calculus. Stud. Math. **136**, 99–119 (1999)
20. Taylor, J.L.: A general framework for a multi-operator functional calculus. Adv. Math. **9**, 183–252 (1972)

Regular Composition for Slice-Regular Functions of Quaternionic Variable

Fabio Vlacci

Abstract In general (see e.g. Cartan in Elementary Theory of Analytic Functions of One or Several Complex Variables, 1963), given two (formal) power series $g(x) = b_0 + xb_1 + \cdots + x^n b_n + \cdots$ and $f(x) = xa_1 + \cdots + x^n a_n + \cdots$ (with $a_0 = f(0) = 0$) it is well known that the composition of g with f, in symbols $g(f(x))$, is a formal power series when coefficients a_j and b_k are taken in a commutative field. Furthermore, if the constant term a_0 of the power series f is not 0, the existence of the composition $g(f(x))$ has been an open problem for many years and only recently has received some partial answers (see Gan and Knox in Int. J. Math. Math. Sci. 30:761–770, 2002). The notion of slice-regularity, recently introduced by Gentili and Struppa (Adv. Math. 216:279–301, 2007), for functions in the non-commutative division algebra $\mathbb{H}$ of quaternions guarantees their quaternionic analyticity but the non–commutativity of the product in $\mathbb{H}$ requires special attention even to define their multiplication (see also Gentili and Stoppato in Michigan Math. J. 56:655–667, 2008). In this paper we face the problem of defining the (slice-regular) composition $g \odot f$ of two slice-regular functions f, g; this turns out to be defined as an extension of the standard composition $g \circ f$ of functions in a non-commutative setting which takes into account a non-commutative version of Bell polynomials and a generalization of the Faà di Bruno Formula.

1 Introduction to Slice-Regularity in $\mathbb{H}$

We recall that the algebra of quaternions $\mathbb{H}$ consists of numbers $x_0 + ix_1 + jx_2 + kx_3$ where x_l is real ($l = 0, \ldots, 3$), and i, j, k, are imaginary units (i.e. their square equals -1) such that $ij = -ji = k$, $jk = -kj = i$, and $ki = -ik = j$. In this way, $\mathbb{H}$ can be considered as a vector space over the real numbers of dimension 4. Given a generic element $q = x_0 + ix_1 + jx_2 + kx_3$ of $\mathbb{H}$ we define in a natural fashion its conjugate $\overline{q} = x_0 - ix_1 - jx_2 - kx_3$, and its square norm $|q|^2 = q\overline{q} = \sum_{k\geq 0}^{3} x_k^2$. The set $\mathbb{S} = \{q \in \mathbb{H} : q^2 = -1\}$ will be referred to as the *sphere of imaginary units of* $\mathbb{H}$.

F. Vlacci (✉)
Department of Mathematics "U. Dini", University of Florence, Viale Morgagni 67/A, 50134 Florence, Italy
e-mail: vlacci@math.unifi.it

G. Gentili et al. (eds.), *Advances in Hypercomplex Analysis*, Springer INdAM Series 1, DOI 10.1007/978-88-470-2445-8_9,

The following is immediate and yet important

Proposition 1 *For any non-real quaternion w, there exist, and are unique, $x, y \in \mathbb{R}$ with $y > 0$, and an imaginary unit I_w such that $w = x + yI_w$.*

Definition 1 Given any imaginary unit I, the set $\mathbb{R} + \mathbb{R}I$ will be denoted by L_I.

Notice that after identifying the imaginary unit I_w in $\mathbb{H}$ with the imaginary unit i of $\mathbb{C}$, the set L_{I_w} may be considered as a complex plane in $\mathbb{H}$ passing through 0, 1 and w. In this way, $\mathbb{H}$ can be obtained as an infinite union of complex planes (which will be also called *slices*).

Definition 2 If Ω is a domain in $\mathbb{H}$, a real differentiable function $f : \Omega \to \mathbb{H}$ is said to be *slice-regular* if, for every $I \in \mathbb{S}_{\mathbb{K}}$, its restriction f_I to the complex line $L_I = \mathbb{R} + \mathbb{R}I$ passing through the origin and containing 1 and I is holomorphic on $\Omega \cap L_I$.

In particular any slice-regular function is C^∞ in $\mathbb{B}(0, R)$.[1] Actually, something more is proven in [5]:

Theorem 1 *A function $f : B = B(0, R) \to \mathbb{K}$ is regular if, and only if, it has a series expansion of the form*

$$f(q) = \sum_{n=0}^{+\infty} q^n \frac{1}{n!} \frac{\partial_C^n f}{\partial x^n}(0)$$

converging in B. In particular if f is regular then it is C^∞ in B.

For an introductory survey on slice-regular functions we refer the interested reader to [7].

Let $f(q) = \sum_{n=0}^{+\infty} q^n a_n$ be a given slice-regular function whose associated quaternionic power series has radius of convergence $R > 0$ and consider the quaternionic power series with real coefficients $\sum_{n=0}^{+\infty} q^n |a_n|$; it also has radius of convergence R, because, according to Hadamard's Formula (see e.g. [1]),

$$R = \begin{cases} 1/\limsup_{n\to\infty} |a_n|^{1/n} & \text{if the lim sup is finite and different from 0} \\ 0 & \text{if } \limsup_{n\to\infty} |a_n|^{1/n} = +\infty \\ +\infty & \text{if } \limsup_{n\to\infty} |a_n|^{1/n} = 0. \end{cases}$$

Therefore the function

$$z \mapsto \sum_{n=0}^{+\infty} q^n |a_n|$$

[1] This smoothness is considered with respect to the so called *Cullen derivative* ∂_C (see also [2, 5]) which is well-defined for slice-regular functions as follows: if $f(q) = \sum_{n=0}^{+\infty} q^n a_n$ then $\partial_C f(q) := \sum_{n=1}^{+\infty} nq^{n-1} a_n$.

(denoted by $f_{abs}(z)$) is a slice-regular function in $B(0, R)$ with the property that for any $I \in \mathbb{S}$, we have $f_{abs}(L_I) \subset L_I$.

2 Non-commutative Bell Polynomials and Slice-Regular Composition

Let $k = (k_1, k_2, \dots, k_n) \in \mathbb{N}^n$ be a given multiindex; we set

$$k! := k_1!k_2! \cdots k_n!$$
$$|k| := k_1 + k_2 + \cdots + k_n$$
$$\|k\| := k_1 + 2k_2 + \cdots + nk_n.$$

Then for $n \geq 1$, we define

$$B_n(y_1, \dots, y_n) := \sum_{\|k\|=n} \frac{n!}{k!}\left(\frac{y_1}{1!}\right)^{k_1} \cdot \left(\frac{y_2}{2!}\right)^{k_2} \cdots \left(\frac{y_n}{n!}\right)^{k_n}$$

where y_j are elements of a commutative algebra; we observe that these polynomials satisfy the recursive equation

$$B_{n+1} = \sum_{k=0}^{n} \binom{n}{k} B_{n-k} y_{k+1} \tag{1}$$

with the initial condition $B_0 = 1$. These polynomials are called *Bell polynomials*. If one decomposes B_n into its homogeneous[2] parts $B_{n,d}$ (with $d = 1, 2, \dots, n$), one can write

$$B_n = \sum_{d=1}^{n} B_{n,d}$$

and obtain

$$B_{n,d}(y_1, \dots, y_n) := \sum_{\|k\|=n,\, |k|=d} \frac{n!}{k!}\left(\frac{y_1}{1!}\right)^{k_1} \cdot \left(\frac{y_2}{2!}\right)^{k_2} \cdots \left(\frac{y_n}{n!}\right)^{k_n}.$$

The homogeneous polynomials $B_{n,d}$ appear in the expression of the n-th derivative of the chain rule, namely

Proposition 2 (Faà di Bruno Formula) *If $h = g \circ f$, then*

$$h^{(n)}(x) = \sum_{d=1}^{n} B_{n,d}\big(f'(x), f''(x), \dots, f^{(n)}(x)\big) \cdot g^{(d)}\big(f(x)\big). \tag{2}$$

[2] A polynomial $B_{n,d}$ is *homogeneous* of degree d if

$$B_{n,d}(\lambda y_1, \lambda y_2, \dots, \lambda y_n) = \lambda^d B_{n,d}(y_1, y_2, \dots, y_n)$$

for any λ.

In [8], the authors extend the notion of Bell polynomials to the setting of a non-commutative algebra $\mathscr{A}$ with unit and obtain their explicit expressions, namely

$$\widetilde{B}_{n,d} := \sum_{n_2,\dots,n_d=1}^{n} \binom{n-1}{n_2}\binom{n_2-1}{n_3}\cdots\binom{n_{d-1}-1}{n_d} y_{n_d} y_{n_{d-1}-n_d}\cdots y_{n_2-n_3} y_{n-n_2}$$

for $n \geq d \geq 2$ and $\widetilde{B}_0 = 1$, $\widetilde{B}_1 = y_1$. It is then natural to consider

$$\widetilde{B}_n = \sum_{d=1}^{n} \widetilde{B}_{n,d}.$$

These polynomials also satisfy the analogous of Eq. (1); in particular, because of the non-commutativity of multiplication, if instead of

$$\widetilde{B}_{n+1} = \sum_{k=0}^{n} \binom{n}{k} \widetilde{B}_{n-k} y_{k+1} \tag{3}$$

(with the initial condition $\widetilde{B}_0 = 1$) one considers the condition

$$\widetilde{B}_{n+1} = \sum_{k=0}^{n} \binom{n}{k} y_{k+1} \widetilde{B}_{n-k} \tag{4}$$

then

$$\widetilde{B}_{n,d} := \sum_{n_2,\dots,n_d=1}^{n} \binom{n-1}{n_2}\binom{n_2-1}{n_3}\cdots\binom{n_{d-1}-1}{n_d} y_{n-n_2} y_{n_2-n_3}\cdots y_{n_{d-1}-n_d} y_{n_d}$$

for $n \geq d \geq 2$ and $\widetilde{B}_0 = 1$, $\widetilde{B}_1 = y_1$. Moreover the expressions of $\widetilde{B}_{n,d}$ can be also inductively obtained by means of a derivation[3] D plus the multiplication from the left of an element of $\mathscr{A}$. In fact, if we adopt the notation $D(y) = y'$ and $D(y^{(n-1)}) = y^{(n)}$, the (non-commutative) Bell polynomials which satisfy (3) are uniquely determined by

$$\widetilde{B}_{n+1}\big(y, y', y'', \dots, y^{(n)}\big) = (D + y)\widetilde{B}_n\big(y, y', y'', \dots, y^{(n-1)}\big), \qquad \widetilde{B}_0 = 1,$$

where

$$(D + y)\widetilde{B} := D(\widetilde{B}) + yB;$$

[3]A derivation is an endomorphism

$$D : \mathscr{A} \to \mathscr{A}$$

of an associative (generally non-commutative) algebra $\mathscr{A}$ with unit 1, such that $D(y_1 \cdot y_2) = D(y_1) \cdot y_2 + y_1 \cdot D(y_2)$. In particular $D(1) = 0$.

for (non-commutative) Bell polynomials which satisfy (4) one has to consider the action of $(D+y)$ on the right. Thus one immediately observes that $\widetilde{B}_{n,1}(y_1, y_2, \ldots, y_n) = y_n$ and $\widetilde{B}_{n,n}(y_1, y_2, \ldots, y_n) = y_1^n$; more in general

$$\begin{aligned}
\widetilde{B}_0 &= 1 \\
\widetilde{B}_1(y_1) &= y_1 \\
\widetilde{B}_2(y_1, y_2) &= y_2 + y_1^2 \\
\widetilde{B}_3(y_1, y_2, y_3) &= y_3 + \underbrace{y_2 y_1 + 2 y_1 y_2}_{\widetilde{B}_{3,2}(y_1, y_2, y_3)} + y_1^3 \\
\widetilde{B}_4(y_1, y_2, y_3, y_4) &= y_4 + \underbrace{y_3 y_1 + 3 y_2^2 + 3 y_1 y_3}_{\widetilde{B}_{4,2}(y_1, y_2, y_3)} + \underbrace{y_2 y_1^2 + 2 y_1 y_2 y_3 + 3 y_1^2 y_2}_{\widetilde{B}_{4,3}(y_1, y_2, y_3, y_4)} + y_1^4 \\
&\vdots
\end{aligned}$$

In any case, it turns out that

$$\begin{aligned}
\left|\widetilde{B}_n(y_1, \ldots, y_n)\right| &\le \sum_{d=1}^{n} \left|\widetilde{B}_{n,d}(y_1, \ldots, y_n)\right| \\
&\le \sum_{d=1}^{n} B_{n,d}\big(|y_1|, \ldots, |y_n|\big) = B_n\big(|y_1|, \ldots, |y_n|\big).
\end{aligned} \tag{5}$$

Assume now two slice-regular functions $f(q) = \sum_{n=0}^{+\infty} q^n a_n$ and $g(q) = \sum_{n=0}^{+\infty} q^n b_n$ are given. Since $f(0) = a_0$ and $\partial_C^n f(0) = n! a_n$, in analogy to (2), we define $g^{\odot} f(q)$ to be the slice regular function

$$g^{\odot} f(q) := \sum_{n=0}^{+\infty} q^n c_n$$

whose coefficients are given by

$$c_0 = g(a_0), \qquad c_n = \frac{1}{n!} \sum_{d=1}^{n} \widetilde{B}_{n,d}(a_1, 2! a_2, \ldots, n! a_n) \cdot g^{(d)}(a_0), \quad n \ge 1, \tag{6}$$

with $\widetilde{B}_{n,d}$ the corresponding non-commutative Bell polynomial. In particular, if $f(0) = 0 = a_0$, then we have

$$c_n = \frac{1}{n!} \sum_{d=1}^{n} d! \widetilde{B}_{n,d}(a_1, 2! a_2, \ldots, n! a_n) \cdot b_d.$$

Clearly, from (5), for $n \ge 1$, one has

$$\begin{aligned}
|c_n| &\le \frac{1}{n!} \sum_{d=1}^{n} \left|\widetilde{B}_{n,d}(a_1, 2! a_2, \ldots, n! a_n)\right| \cdot \left|g^{(d)}(a_0)\right| \\
&\le \frac{1}{n!} \sum_{d=1}^{n} B_{n,d}\big(|a_1|, 2!|a_2|, \ldots, n!|a_n|\big) \cdot \left|g^{(d)}(a_0)\right|
\end{aligned} \tag{7}$$

and the last sum of the right-hand side in the previous inequality corresponds to the coefficients of the power expansion of $g_{abs} \circ f_{abs}$. In other words, we have the following

Proposition 3 *Given two slice-regular functions* $f(q) = \sum_{n=0}^{+\infty} q^n a_n$ *and* $g(q) = \sum_{n=0}^{+\infty} q^n b_n$ *in* $\mathbb{H}$, *if the composition* $g_{abs} \circ f_{abs}$ *of the corresponding associated abs-power series exists and have radius of convergence* R *then it is possible to define the slice-regular function* $g^{\odot} f$ *for* q *such that* $|q| < R$ *in terms of the power series*

$$g^{\odot} f(q) = \sum_{n=0}^{+\infty} q^n c_n$$

where the coefficients are given in (6).

Proof Indeed, the coefficients given in (6) satisfy inequality (7) and this guarantees the convergence of the series $\sum_{n=0}^{+\infty} q^n c_n$ for any q such that $|q| < R$. Finally, because of Theorem 1, the function

$$q \mapsto \sum_{n=0}^{+\infty} q^n c_n$$

is slice-regular. □

Remark 1 The functions f_{abs} and g_{abs} have power series expansions whose coefficients are real numbers, so all the classical results in [1] (and more recent ones in [3]) which provide sufficient conditions for the existence of the composition of formal power series with coefficients in a commutative field apply. In particular the composition $g_{abs} \circ f_{abs}$ exists if $a_0 = 0$.

Remark 2 Since apparently associativity of the product is never applied, a similar result should hold true also for slice-regular functions of octonionic variable (see [6]).

Acknowledgements The author has been partially supported by Progetto MIUR di Rilevante Interesse Nazionale *Proprietà geometriche delle varietà reali e complesse* and by G.N.S.A.G.A (gruppo I.N.d.A.M).

References

1. Cartan, H.: Elementary Theory of Analytic Functions of One or Several Complex Variables. Addison-Wesley, Reading (1963)
2. Cullen, C.G.: An integral theorem for analytic intrinsic functions on quaternions. Duke Math. J. **32**, 139–148 (1965)
3. Gan, X.-X., Knox, N.: On composition of formal power series. Int. J. Math. Math. Sci. **30**, 761–770 (2002)

4. Gentili, G., Stoppato, C.: Zeros of regular functions and polynomials of a quaternionic variable. Mich. Math. J. **56**, 655–667 (2008)
5. Gentili, G., Struppa, D.C.: A new theory of regular functions of a quaternionic variable. Adv. Math. **216**, 279–301 (2007)
6. Gentili, G., Struppa, D.C.: Regular functions on the space of Cayley numbers. Rocky Mt. J. Math. **40**(1), 225–241 (2010)
7. Gentili, G., Stoppato, C., Struppa, D.C., Vlacci, F.: Recent developments for regular functions of a hypercomplex variable. In: Hypercomplex Analysis. Trends in Mathematics, pp. 165–186. Birkhäuser, Basel (2009)
8. Schimming, R., Zagloul Rida, S.: Non commutative Bell polynomials. Int. J. Algebra Comput. **6**(5), 635–644 (1996)

MIX
Papier aus verantwortungsvollen Quellen
Paper from responsible sources
FSC® C105338

If you have any concerns about our products,
you can contact us on
ProductSafety@springernature.com

In case Publisher is established outside the EU,
the EU authorized representative is:
Springer Nature Customer Service Center GmbH
Europaplatz 3, 69115 Heidelberg, Germany

Printed by Libri Plureos GmbH
in Hamburg, Germany